PRINT ORDER PROCESSING

Hugh M Speirs MA BA FIMgt FIOP MIQA LCGI

BRITISH PRINTING INDUSTRIES FEDERATION
11 Bedford Row, London WC1R 4DX

Acknowledgements

The British Printing Industries Federation wishes to thank all who have provided assistance in any way with the production of this book, including MIS companies, printing equipment and services suppliers, who have provided information, data, photographs and transparencies for reproduction.

ISBN 0 851 68197 2

Designed and typeset by Hugh and Jonathan Speirs using Aldus PageMaker 5.0, with text set in Palatino 10 on 12 pt, headings in Helvetica Bold. Cover printed in four colour process on Invercote G 240 g/m^2 as outer spread sheetwork and then OPP laminated.

Text printed in black ink, sheet work on Fineblade Cartridge 115 g/m^2 in 32 page sections as i–10 and 11–26; 27–42 and 43–58; 59–74 and 75–90; 91–106 and 107–122; 155–170 and 171–186; 187–202 and 203–218; 219–234 and 235–250; 251–266 and 267–282 plus 2 up x 16 page section 123–138 and 2 x 8 page sections in 4 colour process 139–146 and 147–154.

Folded, gathered sections, thread sewn and soft back bound, trimmed flush.

Printed and bound by Page Bros (Norwich) Ltd

British Printing Industries Federation, 11 Bedford Row, London WC1R 4DX

Telephone 071 242 6904, Facsimile 071 405 7784

Contents

Introduction

Print Order Processing addresses the considerable changes which have taken place in the administrative procedures and processing of print-related products and services in recent years.

The role of the printing industry in the 1990s if it is to succeed and advance, like so many other industries, needs to be flexible and responsive to its customers needs and to adapt to a seemingly ever-changing business environment. Technology, or more correctly the application of new technology, allied to more-and-more powerful, open and accessible computer facilities has brought printing into the much wider arena of multi-media, digital information and processing.

Up until recent times it has been the norm for printing companies to service customer requirements by maintaining largely separate and discrete job functions and roles. This often resulted in a print customer being in contact with several different individuals within the one organisation all handling the same job at different phases of the job cycle. Depending on the size and structure of the organisation this could be even more expanded and fragmented, with smaller firm's personnel, by necessity, performing several functions. In order to improve customer service this approach is now changing, with printing organisations requiring staff with a much wider range of skills that can be responsive and adaptive to change.

This development has been accelerated by the use of computer software packages. These include sales order processing, estimating, costing, production control, stock control etc. possibly initially as a stand-alone estimating module but often leading to a management information system (MIS) based on an integrated management approach.

Print Order Processing is aimed primarily at the growing role (or roles) of the *account executive*, also known under the job titles of *customer contact*, *customer service* and *customer support representatives* reflecting this customer-centred role. The specialist roles of the 'dedicated' sales representative, estimator or progress chaser/production controller are still required in most medium- to larger-sized printing companies; however they and many others working within a print-ordering environment should also benefit from the contents of this book, as it considers the wide range of knowledge and skills required to process print orders successfully and profitably.

This book underpins the British Printing Industries Federation's annual examination *Print Order Processing* and should therefore be of particular value to students pursuing this qualification through the British Printing Industries Federation correspondence course, printing college or private study.

Further publications in this series are *Introduction to Printing Technology, Estimating for Printers* and *Getting the Measure of your Business.* The reader is particularly directed to the *Introduction to Printing Technology* book for a sound grounding in the technical aspects of printing. This *POP* book assumes at least a basic knowledge of the printing processes and their applications. Although there are areas of overlap with the IPT book, this is intended for reasons of reinforcement, as the main approach of this book is to bring together the application of technical knowledge, working practices and procedures into the desired skills required for effectively servicing a print customer's needs and requirements.

The format has been chosen to be 'reader-friendly' with information supplied in easily accessible chunks of text and graphics responding to its role as a general reference book and as a handbook for students. Chapter overviews have been included throughout the book to inform the reader of the overall content of a chapter and its relationship to other matter in the book, providing a useful point of reference. The *Glossary, Bibliography* and *Index* provide more in-depth details and referencing for particular areas and topics.

1 The printing industry today

The printing industry, as with so many other industries UK and worldwide, has had to adapt to massive changes in the 1980s and 1990s. Printing, for so long a relatively 'closed' craft-based industry, has become part of the wider multi-media/communications industries where publications, documents and records can be reproduced in so many different ways apart from the printed form.

To respond to these challenges the printing and publishing industries have responded by embracing new technology; developing DTP (*desktop publishing*) way beyond its original concept to generation and manipulation of data in increasingly digital rather than analogue form.

The structure and organisation of the present UK printing industry has evolved from very long traditions of guilds, crafts, master printers and apprenticeships, adapting and adjusting to changing circumstances.

This chapter reflects the well-established structure still in place in the industry as well as the many changes which printing has undergone, for example, the move to a more capital-intensive base.

The specific areas covered are as follows: structure of the printing industry - printing in the EC; UK printing industry's trading position; developing changes in the printing industry; printing companies and their customer base - general printers, specialist printers; employers, employees and associated organisations in the printing industry; education and training in the printing industry - entry routes, skills training schemes/occupational standard of competence.

Structure of the printing industry

The printing industry in the United Kingdom has for some years been among the top 10 manufacturing industries - as at 1993 it was the seventh largest.

In that year it is estimated that the industry employed approximately 161 000 people (excluding newspaper printing and publishing) - see *Figure 1.1* - in over 10 000 printing companies. In addition, there were approximately 5000 'print shop/quick printers' and 5000 'inplant' printing units.

Printing is mainly a *bespoke industry* producing products to a customer's specific requirements - eg annual accounts, brochures, cartons, labels and business forms - the main exceptions are combined publishers/ printers such as newspapers and books. The printing industry is made up mostly of small companies. It is also often referred to as 'the barometer of the economy' as it appears to reflect the general economic climate at any time - in other words a buoyant printing industry often reflects a buoyant economy and vice versa.

Figure 1.1 is taken from the BPIF membership database at September 1993. The figures show a breakdown of British Printing Industries Federation members - with print production facilities - in terms of numbers of employees per company and numbers of companies in each category. In addition to the figures shown there are other BPIF member companies - without print production facilities - consisting of group headquarters, sales and administrative offices etc.

This table highlights the fact that the printing industry contains a large number of very small companies. However, the small companies with up to 24 employees employ only 26.2% of the total number employed. The majority of 54.6% are employed in companies with between 25 and 249 employees - medium to large-sized companies - while the remainder are in very large companies.

Printing serves all sectors of the economy including government, financial services, distributive services and manufacturing industry. Demand for its products is essentially derived from the level of activity in the economy at large and although little of its output is sold directly for personal use, consumers' expenditure has a significant influence on the health of the printing industry.

Company size by number of full-time employees	Total number of companies	% of full-time labour force
1 - 9	975	7.5
10 - 24	922	18.7
25 - 49	401	16.5
50 - 99	214	17.6
100 - 149	74	10.2
150 - 249	50	10.3
250 - 499	26	12.2
500+	9	7.0
Total	**2671**	**100.0**

Figure 1.1: Breakdown of full-time employees in BPIF companies taken from the BPIF membership database at September 1993

According to *Pira International's 'UK printing and publishing statistics 1993'* in 1989 the number of small businesses in the UK print *and* publishing industries combined was 16 200; while large companies with more than 2000 employees had contracted from the previous figure to only eight. The combined workforce for UK printing and publishing was given as 312 000.

It is estimated that the UK printing industry had an annual turnover of £8.6 billion in 1991. *Figure 1.2* shows the printing industry sales by larger manufacturers recorded in 1991 as £5.9 billion: the remaining £2.7 million balance coming from smaller manufacturers in the industry. The structure of the printing industry reflects the diversity and fragmented nature of its market with a few integrated groups, a number of medium-sized companies which tend to specialise, and a vast army of small firms which usually cater for mainly local markets.

Classification	**£ million**
General printing	2400
Book printing	260
Periodicals printing	400
Contract printing of newspapers and colour supplements	290
Packaging	1300
Stationery	1030
Ancillary services	220
Total	**5900**

Figure 1.2: Printing industry sales by larger manufacturers in £ millions (at 1991)

Printing in the EC (European Community)

Printing companies in the EC employ approximately 670 000 people with an overall turnover of over 50 billion ECU - 85% of the companies employing fewer than 20 people. The remaining 15% is made up mainly of companies employing between 20 and 500 employees with less than 0.5% of companies having over 500 employees - *source, Intergraf Eurostat from figures published 1993.*

UK printing industry's trading position

Competition from imports mainly affects book printing (about 25% of which is imported), long-run advertising material (such as travel brochures and mail order catalogues) and cartons. Exports are dominated by publications and security printing. A high proportion of trade in publications represents sales by publishers and is only indirect exporting for the printing industry. In recent years, however, there has been an encouraging growth in direct exporting by UK printers. The balance of trade in printed matter remained positive throughout the 1980s and in 1992 it rose to £151.6 million compared with £141 million in 1990.

Developing changes in the printing industry

In the 1980s and 1990s printing became more capital-intensive compared to its predominately labour-intensive, mainly craft-based structure - see *Figure 1.3* below.

UK printing and publishing plant and machinery expenditure

Year	£ million
1983	242
1984	344
1985	434
1986	482
1987	560
1988	722
1989	855
1990	753

Figure 1.3: Capital investment figures for years 1983 to 1990 taken from Pira International's 'UK printing and publishing industry statistics 1993'

Up to 50% of the labour force is not directly or exclusively employed in production, but employed in 'support' roles such as management, administration, sales and customer service staff. Although job roles and working practices have changed considerably during the last decade in the printing industry, the labour force still remains predominantly male, especially in the production areas.

Classification	Number of employees
General printing	75 500
Book printing	10 000
Periodicals and colour supplements	10 600
Cartons	16 600
Flexible packaging	8 200
Stationery	27 200
Ancillary services	12 900
Total	**161 000**

Figure 1.4: Breakdown of labour force per sector taken from the BPIF 'Information File', published September 1993

Printing companies and their customer base

Every town of a reasonable size in Great Britain is served by a number of small to medium-sized printers, established originally to meet the demand for general printing in the locality. Larger cities and conurbations, because of their large and wide diversity of businesses and interests, often lead to large-sized and/or specialist printing companies to meet those needs.

As printing is a service industry, the initial tendency was for printing companies to set up in areas where there was local industry; but today even established 'local' printers supply a much wider geographical area than their own locality. Following developments of new and expanding industries, certain areas have grown up in response as sizeable printing centres undertaking work well beyond the local requirements.

Customers ordering print are as wide ranging and varied as the printing companies serving them - from the one-person business to multi-national organisations in all service and product areas.

General printers

Printing companies serving local communities and business sectors generally fall into the sector known as *general* or *jobbing printers*, reflecting the needs of their customer base in the range of printed work they undertake.

The product range of general printers is wide and varied and includes leaflets, brochures, booklets, folders, general printed stationery, timetables, posters, printed envelopes and cards etc.

The largest sector in the printing industry is in *general printing*. Reference to *Figures 1.2* and *1.4* illustrate the point that general or jobbing printers account for more than 40% of the turnover and employment within the industry.

Specialist printers

Many printing companies and print groups have begun to develop and grow as *specialist printers* in a particular narrow band of products in an attempt to build up expertise and gain the maximum return from a narrower market segment than just general commercial print.

Examples of specialist printers' work includes the production of books, periodicals, cartons and packaging, labels, continuous business forms and direct marketing.

Specialist printers tend to have a much more focused and targeted customer base, offering a narrower printed product range, often of fairly high print runs, so encouraging investment in specialist equipment.

The customer base for *book production and periodicals* is predominantly publishers, especially for mass-circulation items. Some short-run, special-interest publications, in addition to being generated by publishers, are also prepared and controlled by sponsoring organisations and societies.

Cartons and packaging cover a wide range of items aimed at storing, identifying, protecting, displaying and enhancing the 'host' product, which covers all sectors of industry - eg food, pharmaceuticals, automotive, lighting, clothing, fancy goods, footwear, liquids etc. Print buying customers in this area are either from advertising/graphic design agencies or direct from the manufacturer or supplier.

Labels cover sheet and roll products, again across a vast and wide range similar to that outlined above for cartons and packaging. Here the customer base is direct from the manufacturer and supplier, or advertising/graphic design agencies.

Continuous business forms is unique in comparison with other print sectors in that it has a high element of trade or print brokering where printers not specialist in this area order their clients' continuous forms requirements through trade printers. In addition, customers across all industries order forms and stationery direct from continuous business forms printers.

Direct marketing covers from a basic product reply card to a highly innovative series of complex items within one mailer product. The main customer base includes the following sectors: financial, mail order, charity, publications and music, holidays and travel, utilities, luxury, motor and clothes etc.

Two further areas of specialist printing are *inplants* and *print shops*. These have developed over the last 30 years in the UK in response to an identified opportunity or need in a market sector, either not covered or not adequately covered by traditional printing companies, in terms of cost and service.

Inplant units, being sited in the host organisation's premises, or within close proximity, have the main objective of supplying at least the majority of day-to-day print requirements in the short-run single and 'spot' colour areas. Some inplants are large in their own right, with up to multi-colour large sheet and web presses.

In recent years it has become fairly common for a *facilities management group* to take over and manage the inplant equipment, either continuing to run the unit from its present site or from a remote, central supply unit.

Print shops, also known as instant or quick printers, were to some extent set up to compete with the traditional local 'High Street' jobbing printer with a general stationery shop at the front and print unit at the back.

Print shops, although starting from humble beginnings, have grown tremendously, introducing a professional, 'quick print' service and response. The growth of electronic printing has helped print shops respond in a positive way, resulting in a much higher profile, quick response personal print service. In an attempt to gain 'economies of scale', established profile and increased publicity, franchises have become popular with print shops, leading to service and product identification in a particular market sector.

The size of a printing establishment, its equipment and processes, along with the skill of its workforce, determines the class and quality of work it is best able to undertake. Few companies, apart from the very largest or as members of a printing group, are equipped to cover in-house all the stages printed matter can go through.

Very few companies combine different processes - for example, offset lithography, gravure, flexography, screen - on one site, other than possibly in the specialist area of narrow-reel printing of labels and packaging, which can combine printing units of all major printing processes on one multicolour press.

Most printing factories specialise in only one of the major processes, normally either in sheet- or web-form.

Although most printing companies try to produce as much work as possible on their own equipment, it is uneconomic to install plant which is not going to be used very often. Trade houses have therefore sprung up to meet the demand for particular specialist services. These include trade typesetters/bureaux, colour scanning/repro services, trade printing and finishing, perfect binding, laminating and die cutting.

In recent years it has become relatively common for some firms, especially large companies, to relocate outside built-up areas such as towns and cities, moving to green-field sites, business or industrial parks. The move often coincides with a major change in equipment, processes or working practices - eg investment in new technology with greater use of automation and computer links. This can result in streamlining of production and administrative procedures.

In time, closer and closer *printer-customer links* are likely to develop in all areas of print-related services and products; customers will increasingly decide how involved they wish to become in generating, manipulating and managing the finished printed product.

Employers', employees' and associated organisations in the printing industry

The printing industry has had employers' and employees' organisations ever since the Elizabethan era. They were initially confined to the London area for centuries but spread throughout the British Isles after the industrial revolution. The twentieth century saw a consolidation of national organisations, a process which started in the 1890s.

BPIF *(British Printing Industries Federation)*

The BPIF can be broadly described as the national association of employers in the general printing industry. It is an independent association owned and directed by its members whose main purpose is to encourage the efficiency and profitability of the industry.

Founded in 1900, the BPIF today represents almost 3000 printing, typesetting, platemaking, finishing and bookbinding companies of all sizes.

It maintains close relations with the Scottish Print Employers' Federation (SPEF), the Irish Master Printers' Association, the Irish Printing Federation and affiliated printing employers' federations overseas. In addition, it plays an active part in Intergraf, the international printing employers' organisation based in Brussels.

NS *(Newspaper Society)*

The Newspaper Society represents the interests of regional provincial newspapers and local press covering approximately 1000 daily and weekly newspapers - it advises members on all areas which concern them as newspaper proprietors, except those relating to the actual supply of news.

NPA *(Newspaper Publishers Association)*

The Newspaper Publishers Association is the trade association for the national newspaper industry, covering all the national daily and Sunday publications and London Evening Standard.

GPMU *(Graphical Paper and Media Union)*

The Graphical Paper and Media Union was formed in 1991 when the two previous separate printing trade unions - NGA *(National Graphical Association)* and SOGAT *(Society of Graphical and Allied Trades)* - amalgamated. Prior to the amalgamation, SOGAT was reported to have over 150 000 members and the NGA over 120 000 members. The total GPMU membership recorded at year end September 1992 was just below 270 000 of which just over 82% was male and just under 18% was female.

NUJ *(National Union of Journalists)*

The National Union of Journalists is the major union covering journalists in Great Britain and the Republic of Ireland, with a membership made up of journalists, photographers and creative artists.

PIRA *(Printing Industries Research Association)*

PIRA, the research association for the paper and board, printing and packaging industries, is an autonomous and independent body financed by voluntary subscriptions from firms, consultancy earnings, and Government grants. PIRA members have access to a range of technical advisory and information services.

IOP *(Institute of Printing)*

The Institute of Printing promotes the advancement of the science and art of printing and bookbinding and sets standards for examinations which lead to professional membership of the Institute - courses which lead to membership include the Ordinary National Certificate in Printing and the Higher National Certificate in Printing.

Education and training in the printing industry

The BPIF Industry Training Organisation (ITO) made up of BPIF staff, provides services for the entire printing industry. It was set up in 1982 and provides resources for the British Printing Lead Body (BPLB) and the Printing Industry Qualifications Council (PIQC).

Its mission is: 'To provide the services, products and incentives, which enable all sectors of the industry to attract and develop high quality people at every level, so as to improve the profitability of companies and meet the career needs of individuals'.

Entry routes

There are three possible points of entry for those wishing to follow a career in the printing industry.

First there is the *skilled production worker entrant* who will often follow the appropriate scheme of training agreed between the BPIF and the GPMU.

Secondly there is the *office entrant* in which students may supplement their training by attendance at business and technical colleges or by taking one or more of the correspondence courses organised by the British Printing Industries Federation.

Thirdly for *administrative and management entrants* there are a range of full-time, block-release, day-release and evening courses run at universities, business and printing colleges covering from degree level, BTEC, City and Guilds to local college certificates in a wide range of management, administrative and technical subjects.

Although three separate and discrete groups have been identified, it must be acknowledged that one person could easily pass through all areas during their career in printing - eg starting as a production worker, then moving on to become a technical co-ordinator in the general office and finally a senior management position, either within the same company or different companies.

Skills Training Schemes/Occupational Standard of Competence

The *Printing Skills Training Scheme* is run by the BPIF Training Organisation. It is an integrated programme combining on-the-job skills training with college study leading to a nationally-recognised vocational qualification. In 1993 there were 20 participating colleges covering at least some of the seven subject areas - origination, machine printing, bookbinding, print finishing, carton manufacture, screen printing and reprography; as well as a wide range of basic skills and skills development modules.

Printing in the UK, as with most other industries, is working towards developing occupational standards of competence appropriate to its own requirements.

The BPIF (and SPEF), are developing NVQ or SVQ *(National/Scottish Vocational Qualifications)* in production printing skills at levels 2 and 3, plus administrative and management qualifications.

The BPIF also runs qualification courses, jointly with Henley Management Centre, providing awards at NVQ levels 4 and 5 in management.

2 The administration and processing of printed matter

The printing industry is made up of a multifarious group of companies which differ in terms of size, core business or markets served, organisational structure and development. *Chapter 1* covered the wide range of general and specialist printers addressing different market sectors.

This diversity associated with the printing industry is hardly surprising as the industry ranges from the one-person business up to large printing companies or groups having hundreds and even thousands of employees. There are clearly therefore many different organisational structures which have been devised to meet these varying requirements - the form printing organisations adopt depend very much upon the scope of their activities.

Although acknowledging the differences between the wide spectrum of printing companies operating in the industry, it is still a fact that there is a major common core of functions and activities which must be addressed and performed by small and large organisations alike to run successfully and efficiently.

This chapter will address these by covering the following areas: marketing, sales and administrative functions; relationship between 'outside' sales staff and account executives/customer support services; procedural routines; forms used in the office; forms used in the office and works; works/production departments and organisation charts.

Marketing, sales and administrative functions

Being part of a bespoke industry, printed matter has to be sold before it is produced.

Marketing

The main objective of marketing, as with other management and administrative functions, is to achieve and sustain greater profitability for the organisation. The importance of marketing is emphasised by a quote from Peter Drucker, an outstanding management writer: 'Marketing is so basic that it cannot be considered a separate function.... It is the whole business seen from the point of view of its final result, that is, from the customer's point of view.'

Organisations that accept marketing as one of their central driving forces, set out to identify, obtain and retain a range of targeted customers, while at the same time making the most effective use of their resources.

Marketing is often expressed in terms of the *marketing mix,* which is basically a set of controllable variables which an organisation can adjust in response to customer behaviour.

The marketing mix is often identified according to the four factor classifications of product, price, place and promotion -

Product - what products are to be sold?

Price - at what price are they to be sold?

Place - to whom are they to be sold?

Promotion - how are they to be sold?

The four P's all make their own contribution to the overall success of the marketing approach. Organisations need to ensure they supply the right *product* in the right *place* (at the right time) and at the right *price* - *promotion* ensures that the organisation's goods and services are made known to prospective and existing customers.

If the marketing mix is correct, then the organisation will have a better chance to be successful and remain so. If the products produced are not required by sufficient customers then the organisation will tend to trade unprofitably, often through under-utilised resources; problems also arise if goods are not supplied at the right place - eg distributed to customers branches as pre-arranged. When prices are too low, then a commercial opportunity is being lost as well as reduced profits or even losses; if prices are too high, customer demand will fall, resulting in a reduced customer base and again lower profitability. Although an organisation can build up a level of expertise and reputation as a dependable high-quality printer, promotion in at least some form or other ensures the marketing drive is maintained.

By adopting a marketing approach, a printing organisation changes from the traditional 'order-taker' approach to a *market driven one,* with an informed awareness of its customer base and markets.

It is often assumed that marketing is not appropriate for small-scale organisations such as printers; on reflection this can be seen to be largely unfounded. Smaller organisations generally have fewer resources and the more effectively these can be used, the more successful the organisation can be. If printers can utilise their skills and expertise effectively within the requirements of the market place, then they will be in a stronger position to survive and prosper.

Sales

One of the most important aspects of an organisation adopting a marketing plan and strategy is that the sales personnel are aware of which products and services best suit their organisation and those of their customers.

It is the function of the sales side of a business to make and maintain contact with customers and to create a demand for its products and production capacity by attending to the customer's requirements and enquiries. Many printing companies, in recent years, have refined the area of sales/customer contact into *'outside' contact* and *'inside' contact.*

The 'outside' contact, the sales staff's main function, is to create and attend to existing and potential customers' enquiries by visiting the clients regularly and on request, passing on the requests for quotations and orders to the next person in the enquiry/order processing chain. This can often simply be the estimator, but increasingly is the account executive/customer support services acting as 'inside' contacts, who will be responsible for handling accounts or individual jobs from inception to final despatch and delivery to the customer.

Relationship between 'outside' sales staff and account executives/customer support services

Sometimes difficulty arises in deciding the relative responsibilities between 'outside' and 'inside' customer contact. On the one hand 'outside' sales staff may not be as knowledgeable as customer support services from a technical point of view, since their function requires, rather than technical knowledge and administrative procedures, a pleasing personality, the ability to create confidence, to discuss the proposed job in the customer's own language, and to close a sale.

Account executives and other members of customer support services, on the other hand, must be able to interpret customers' requirements in an analytical manner, translating any particular job into technical terms for all the printing departments concerned.

The 'inside' customer contact staff must not think that because 'outside' sales staff talk in non-technical language, they know less than they do of what the customer wants. Mutual understanding that allows each job to be discussed fully and freely must be evident in both parties, the main purpose being a *correct interpretation* of customers' requirements, and the making of arrangements for each job to be produced in the most expeditious and economic manner.

Administration

The main administration functions in a printing company are estimating, account executives/order control, production control, purchasing, costing and pricing/invoicing.

Some of the above administrative functions, especially estimating and account executives/order control, are often included under the combined title of *'Customer support services'*; with all administrative functions often coming under the title *'Commercial services'*.

Production control is covered in depth in *Chapter 6*, pages 89 to 102.

In larger organisations each of the administrative functions is often separate, requiring the services of several persons, whereas in smaller concerns, some are often grouped together under the responsibility of one person.

Estimating. This entails preparing estimates and, in conjunction with the sales representatives and senior management, preparing and servicing the quotations submitted. The person preparing the estimates has to calculate all the cost elements of the enquiries and prepare draft estimates. This is then checked by the senior estimator and/or by another member of senior management who decides the actual selling price. The final quotation is then prepared, typed out and submitted to the customer for consideration.

Account executive/order control. The account executive prepares complete working specifications and services the routine procedures for handling orders received from customers. This includes issuing appropriate instructions to enable orders to be put in hand, despatch and acknowledgement of orders and proofs and other numerous duties relating to the progress of orders throughout the works.

Purchasing. A function which can be carried out by the account executive or even the estimator when the quotation has been accepted, or by a specialised purchasing department. Efficient purchasing ensures the right material is available at the right time on the best possible terms.

Costing. Costing is concerned initially with finding the costs of running the whole business - offices and works - for a period ahead, usually a year; cost rates are calculated which will be used for estimating, control and pricing purposes.

The BPIF system of management information, which is used by most printers in this country as the basis of their own costing systems, enables businesses to control all aspects of costs and generally to become more efficient and profitable.

This is achieved by the imposition of a close and continuous check on all expenditure, on the use of resources and on the recovery costs. Every expense incurred directly and indirectly in the production of a job is recorded, so that the cost of production may be determined.

Pricing and invoicing. Pricing is the determination of the price to be charged to the customer either at the time a quotation is requested or after a job has been delivered and the cost of its production is known. It is essential that the quotation, if one was prepared, and the cost sheet for each job are compared. The specification in the quotation should agree with the production of the finished job and where variations have taken place, as they often do in production, they must be taken into account when arriving at the price to be charged.

Other additional charges not shown in the quotation, such as author's corrections, overtime premium charges, additional sets of proofs and excess carriage will be shown as separate additional costs. If the job has not been the subject of an enquiry, the completed cost sheet will be passed through for pricing. This will involve scrutinising the costs and fixing the charge in accordance with the policy of the organisation and finally preparing the wording of the invoice for typing. Historical costs from past job records will be used as a basis for establishing prices on the jobs which have been produced previously to the same or similar format.

Procedural routines

It is particularly essential in printing companies where there is so much variety in the work produced and complexity in the production methods, that a structure of well-planned and co-ordinated procedures is introduced in order to maintain the consistent, smooth and efficient running of the organisation.

Organisations which have acquired or are working towards BS5750/ ISO 9000 certification will normally have introduced *'Procedures'* as level 2 of their Quality Management System (level 1 is the *'Company Quality Manual')*. In this context procedures are prepared to clearly state the detailed departmental operations within the quality system, relationship with other departments and processes, as well as referencing related activities to the overall quality plan.

The procedures' main function is to clarify how the quality policy is to be carried out - *who does what, in liaison with whom, and on what authority*. Level 3 of the system is *'Works instructions'* which covers the detail of how operations and processes are to be carried out; finally, level 4 covers *'Documentation'*.

Documentation or forms are distributed to individuals to assist them in their job so that they can execute, check and record operations or material usage against recognised practices and standards. In order to ensure the overall system operates effectively there should be procedures or controls set up which ensure the forms issued are always of the latest approved version and that individuals 'sign on' and 'sign off' in the spaces provided, when they receive and complete the operations required by the document.

The forms used in printing organisations vary considerably, depending upon the size of the establishment, the type or types or work they produce, and many other factors.

With the increasing use of computers in all organisations, see *Chapter 3, Management Information Systems*, it is becoming more and more common for data to be retained in the computer system with hard copy forms generated as and when required from a set format and integrated database. The forms which are reproduced in the following pages indicate minimum requirements and should be regarded only as a guide to the sorts and styles normally used.

Forms used in the office

1) Estimate form

2) Order record

3) Acknowledgement of order

4) Cost sheet

5) Proofs enclosure slip

6) Customer jobs record.

In addition to these six forms, others may be used to record the level of activity over a period and the effect this has on recovery of cost. Such forms, which would allow management to correct adverse trends, might include a *Recorded costs form* and a *Comparison of budgeted and recorded costs*. Forms relevant to production control are covered in *Chapter 6*.

1) Estimate form

Complete and accurate details of the job, the proposed quality and quantity of materials to be used and the method of production are filled in on the form as appropriate. The estimator works out the time required for each operation and enters the cost details. When the summary is completed, a member of management usually reviews the job as a whole, and determines the price to be quoted.

Estimate form

Customer's name	**ESTIMATE NUMBER**
Address	Date
Specification	Enquiry per/date:
	Representative
	Letter/Fax/Telephone/Verbal

					Print quantity				*Print quantity*		
PAPER/BOARD											
					hours	@	£	*p*	*hours*	£	*p*
				PAPER/BOARD SUB TOTAL							
PRE-PRESS		£	*p*	**Proofing**							
Design/Artwork	@			B & W - laser/photocopy/bromide/ozalid							
Imagesetting/DTP	@			Colour - digital/Cromalin/wet							
Camera/Contacting	@			**Pre-press materials**							
Scanner - B & W	@			Bromide/Film/Foils/Plates							
Scanner - Colour	@			**PRE-PRESS TOTAL (B)**							
Paper make-up	@			**PRE-PRESS SUB TOTAL (A) + (B)**							
Film assembly	@										
	@			**MACHINE PRINTING**							
Retouching	@			MR							
Preparing masks	@										
Platemaking	@			Run							
Author's Corrections	@										
	@			Wash-up							
PRE-PRESS TOTAL (A)											
INK				**MACHINE PRINTING SUB TOTAL**							
				INK/VARNISH SUB TOTAL							
Notes/Methods of working				**PRINT FINISHING**							
				PRINT FINISHING SUB TOTAL							
				Outwork							
Prices quoted											
				Material expenses %							
				EXPENSES SUB-TOTAL							
				Mark-up %							
				TOTAL							

Figure 2.1: Estimate form

2) Order record

Every order will be entered onto an order record system, in the form of a bound book, card index, loose leaf folder or on computer. Brief but sufficient details will be entered against each job number, which will thereafter identify the job throughout the various stages in its production. Subsequently, details may be added regarding the despatch and return of proofs and the delivery of the finished work.

Job No.	Date	Customer	Brief details of order	Date delivered	Day book folio
1001					
1002					
1003					
1004					
1005					

Figure 2.2: Order record

3) Acknowledgement of order

Whether an order is received by letter, fax, official order form, or just verbally, it is in the printer's interests to send the customer, as soon as possible, an acknowledgement of his instructions. On the reverse may be printed the *Conditions of contract*, see pages 84 to 87.

Quality Printers Ltd.
Imprint Works, Newtown
London Office: 1 Caxton Buildings WC1

Dear Sirs,

We acknowledge with thanks your instructions of
received on to proceed with the undermentioned work.
Please quote job no. in any correspondence regarding this order.

DESCRIPTION

Yours faithfully,
QUALITY PRINTERS LTD

Conditions - see reverse

Figure 2.3: Acknowledgement of order

4) Cost sheet

This is the form which ensures that all costs incurred are properly recorded against each job. Each morning the costing personnel will receive the daily dockets from the production departments for the previous day. The first task is to record the direct time from the dockets to the appropriate cost sheets.

When a job has been completed all these times are totalled and evaluated at the appropriate hourly cost rates to give the cost of each operation. These costs are totalled for each department and the amounts transferred to the summary. The costs of direct materials and outwork, to which must be added indirect expenses to cover the cost of buying, storing, handling, packing, delivery, selling, administration and so on, are also entered on the cost sheet. The total will then give the cost on which the charge for the job may be based and this can be compared with the estimate, if one has been prepared.

To obviate the time-consuming task of manually costing each job, a MIS can be used.

Basic data taken from the works instruction ticket and daily dockets are entered into the computer for processing the information on each separate job - such data can be entered via a computer terminal or recorded through a shopfloor data collection system, see page 100. The data store of the computer accepts information on a daily basis on each job, calculates costs based on current recovery rates and produces final totals on completion of the job. At the same time it will use the information to produce records of factory activity and analysis of the reasons for inactivity.

5) Proofs enclosure slip

This is a simple form for despatch with proofs to the customer and may be used to make suitable comments where necessary. The customer should be made aware of the correct proof correction marks, preferably with a special leaflet.

6) Customer jobs record

This form is of value when having to estimate for, or put in hand, a reprint of a job. The essential details of the various jobs which have been produced for a particular customer, and their respective job numbers, can be found easily, thus avoiding searching through files of cost sheets.

Cost sheet

JOB NUMBER ____________________

Date ____________________

Customer ____________________

Description of job ____________________

Customer's Order No.	Date	Quantity
Delivery requested	Delivery promised	Initial

	Quantity	Price	Cost £	Cost p	Cost £	Cost p	Charge £	Charge p
Goods for resale								
Direct materials								
Outwork								
Direct expenses								
Sub-total								
Production								
Total								
Direct services								
Total								

Despatched		Invoiced				£	p	£	p
					Total costs				
Date	Quantity	Date	Inv. No.	Initial	Margin				
					Price				
					Special discount				
					Invoice value				
					V.A.T.				

Remarks		Number	Date	Price
	Previous order			
	Estimate			

Figure 2.3a: Cost sheet (front)

Record of production costs

Budget centre	Cost centre	Operation											Totals	Cost rate	Cost £	p
			Date													
			Time													
			Date													
			Time													
			Date													
			Time													
							Total costs ______				Department			£		
			Date													
			Time													
			Date													
			Time													
			Date													
			Time													
							Total costs ______				Department			£		
			Date													
			Time													
			Date													
			Time													
			Date													
			Time													
							Total costs ______				Department			£		
			Date													
			Time													
			Date													
			Time													
			Date													
			Time													
							Total costs ______				Department			£		
			Date													
			Time													
			Date													
			Time													
			Date													
			Time													
							Total costs ______				Department			£		

Figure 2.3b: Cost sheet (reverse)

Quality Printers Ltd.
Imprint Works, Newtown
London Office: 1 Caxton Buildings WC1

____________ 19

To ______________________________________

____

PROOFS

We have pleasure in enclosing proofs of the undermentioned job for your perusal and approval. Please return marked-up copy as: *Proof approved without correction; Proof corrected as marked, proceed after correction; Proof corrected as marked, reproof.*

With compliments,
QUALITY PRINTERS LTD.

Figure 2.4: Proofs enclosure slip

Customer's name

Address

Date	Works	Details of order	Price	Total Value

Figure 2.5: Customer jobs record

Works and production departments

In both small and large organisations the production elements are usually divided according to the processes undertaken, each department or section being under the supervision of an overseer, foreperson or team leader.

Pre-press departments

These cover all areas from initial concept of a printed job up to surface preparation - for example, plates and cylinders. These can include:

Graphic design studio where design concepts and visuals up to final artwork can be prepared using a combination of mainly manual skills, or accessing through a computerised design software package with supporting equipment to produce up to camera-ready copy, mainly by electronic means.

Letter assembly or typesetting operations which cover a very wide range of options, from traditional keyboarding and outputting, using the customer's marked-up copy, through a range of DTP options, word processing disks and data conversion systems to outputting on imagesetting systems.

Graphic reproduction, where finished artwork from the graphic design studio and from the letter assembly department are reproduced mostly on film, ready for planning and platemaking; this will involve mainly the use of cameras and scanners, mono and colour, plus related processing equipment.

Planning and platemaking which covers planning either manually using a montage of line and screen film laid down onto clear foils, or 'automated' planning on an 'electronic' page make-up system. Film exposure to plates and cylinders can again be mainly manually or automatically controlled.

Machine printing departments

These cover the range of printing equipment used in any printing company from small format sheet or label machines to large sheet and complex web-fed machines with on-line finishing facilities, using the offset litho, gravure, screen, flexography or letterpress processes.

Print finishing

This covers a wide range of operations in which printed sheets, processed reels or webs are bound or otherwise finished.

Despatch

This department links with the print finishing department to ensure the finished jobs are packed and despatched to the customers' requirements.

Materials stores

These govern the stock control and issue of raw materials such as paper, board, inks, film, plates and chemicals.

All printed work should be subjected to an inspection before the finishing stages, and the completed work should be examined again before despatch.

The responsibility for quality control usually rests with the quality manager/controller, if such a separate appointment exists within the organisation. Whether it does or not, it must be understood that quality is everyone's responsibility, each person contributing their own vigilance and expertise to ensure any product or service is to the approved specification and quality before being passed onto the next link of the processing chain.

Every department in the printing works will have its own set of records. For example, there will be forms to record materials received and issued, and the movement of particular jobs in and out of each department.

Forms used in the office and works

1) Works instruction ticket
2) Amendments to order
3) Daily docket
4) Material release and cutting slip
5) Stock control records
6) Order form
7) Spoilage ticket
8) Despatch advice.

1) Works instruction ticket

Receipt of the works instruction ticket is recognised as the authority for the work to be put in hand in each department and provides instructions for the guidance of those concerned in its production. It should be a brief yet complete description of the job, and should follow the sequence of operations required as far as possible and give precisely the relevant specifications and instructions for each process or department.

Duplicate copies, having the same job number, are also prepared for the office, for the materials storekeeper, and sometimes for the advance information of each department concerned in the production of the job, with a view to ensuring a minimum of delay at each stage: several copies are essential when work on a job is required to be carried out in more than one department at the same time.

Although the BPIF issues a standard works instruction ticket many companies prefer to use their own specially-designed forms, or adaptations of the BPIF standard to meet their own particular requirements. Most organisations using a MIS will produce the vast majority of their documentation, including the works instruction ticket, from the system.

For the smaller organisations, the BPIF has created a combined form consisting of a cost sheet, works instruction ticket and daily docket which splits into two parts - the *originator's copy* and the *works and costing copy*. The works instruction ticket must provide a clear and precise specification for the job and should always give clear and accurate instructions.

The top part of the form usually gives the customer's name and address, the order number, details of the order, quantity, size and the delivery date. The bottom part normally gives detailed departmental instructions, and may also have spaces or sections for completion by the department concerned showing the quantities and values of ink, varnish and other materials used, progress dates, proof despatches and returns, and the actual job delivery date or dates.

To assist in quality checks and 'traceability' records there are often checklist boxes where operators 'sign-in' and 'sign-off' the work they have completed, including the appropriate dates.

Construction of the works instruction ticket

The works instruction ticket may be produced in one of the following forms: multi-part set of forms; single paper sheet which is affixed to the job envelope; printed envelope; or job card, but with this method it is difficult to make duplicate copies. A printed sheet or card may be circulated in a transparent polythene bag with other relevant material.

Works instruction ticket

JOB NUMBER ____________________

Date ____________________

Customer ____________________

____

Description of job ____________________

____

____

____

Customer's order no.	Date	Quantity
Delivery requested	Delivery promised	Initial

All copy, layouts, specimens, proofs etc, to accompany these instructions through the works. On completion enter job on daily delivery sheet and pass to Cost office with ________ file copies.				**Order received in department**	
				Date	Initials
Paper/Board	Stock reference	Quantity	Requisition to stores		
Design For special instructions, see reverse	Material				
Imagesetting	Proofs	Sent	Returned		
For special instructions, see reverse	1st 2nd				
Graphic reproduction For special instructions, see reverse	Material				
Planning and Platemaking For special instructions, see reverse	Material				
Proofing	Stock	Quantity reference	Requisition to stores		
For special instructions, see reverse	Inks etc.				

Figure 2.6a: Works instruction ticket (front)

Machine printing For special instructions, see below	**Inks etc.**			
Print finishing For special instructions, see below	**Print finishing materials**			
Outwork	**Supplier**	PO No and initial	Date ordered	Date received
For special instructions, see below				
Deliver to	**Carrier**	Quantity	Date despatched	Initials
For special instructions, see below				
Special instructions				

Packing	**Part deliveries**					
	Date	Quantity	Carrier	Date	Quantity	Carrier

Figure 2.6b: Works instruction ticket (reverse)

Methods of completion

Carbonless multi-part sets, which can either be continuous or cut sheet, are made up of the required number of leaves with the top part of the form filled in. Only the top part of the works instruction ticket and the cost sheet are identical and care should be taken to ensure that the second part of the works instruction ticket is not reproduced on the cost sheet.

Distribution

a) The works instruction ticket gives the authority to proceed and is circulated with the job, going first to the department where the initial work will be carried out. Each inter-departmental transfer goes via the production controller and/or customer services department in order to keep them informed of the progress of the job. The only time the works instruction ticket is separated from the job is when the work has been given out to a trade house, and the ticket is then filed under 'outwork' in the production controller's or customer support services' department.
b) The cost sheet is sent to the costing office.
c) The duplicate copies of the works instruction ticket are sent to the overseers of the pre-press, machine printing and print finishing departments, and to the materials storekeeper.
d) Where a combined works instruction ticket, cost sheet and daily docket form is used, one copy is kept in the general office; the other with the cost sheet on the back is circulated with the job.
e) The forms referred to in a) to d) are returned to the general office after completion of the job.

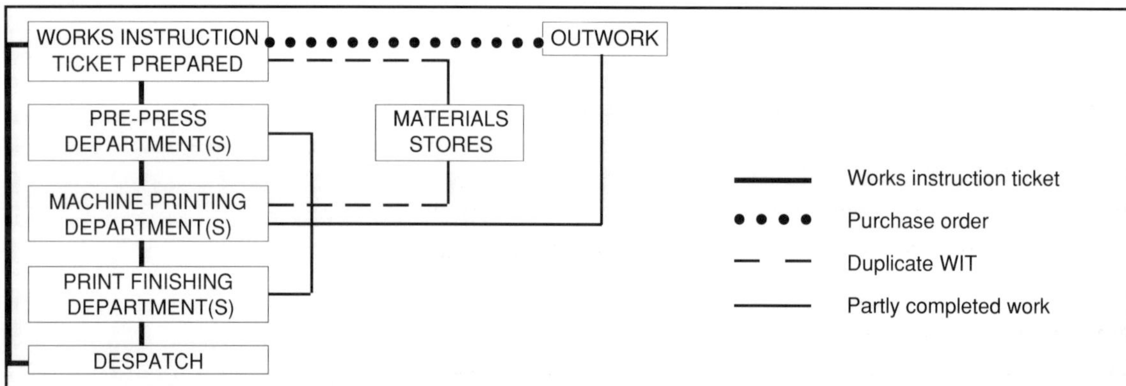

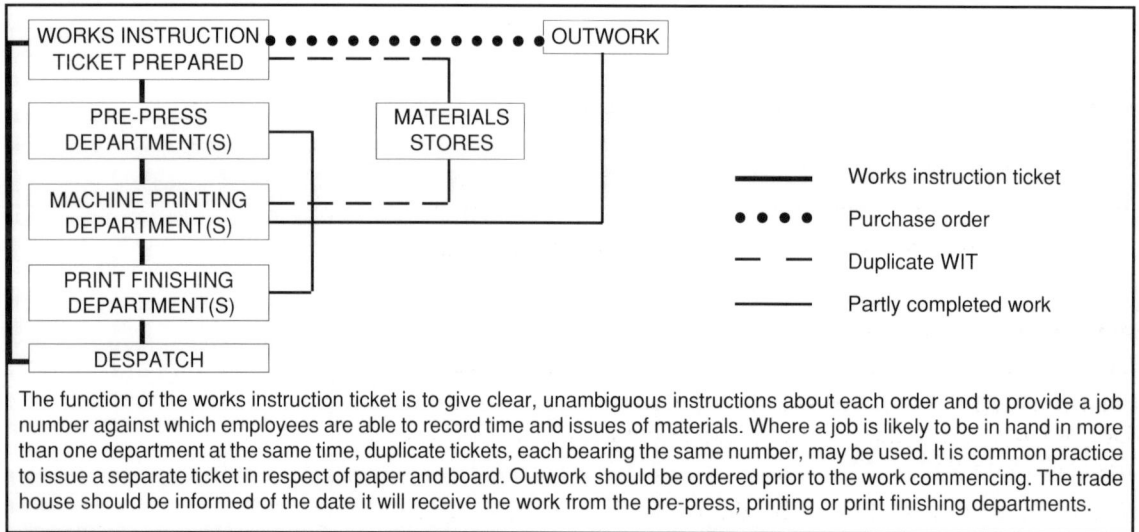

The function of the works instruction ticket is to give clear, unambiguous instructions about each order and to provide a job number against which employees are able to record time and issues of materials. Where a job is likely to be in hand in more than one department at the same time, duplicate tickets, each bearing the same number, may be used. It is common practice to issue a separate ticket in respect of paper and board. Outwork should be ordered prior to the work commencing. The trade house should be informed of the date it will receive the work from the pre-press, printing or print finishing departments.

Figure 2.7: Flow diagram of the works instruction ticket

2) Amendments to order

It sometimes becomes necessary to revise the instructions issued originally on the works instruction ticket, for example, the quantity to be printed may be increased, pagination, colour of ink, or paper and board quality may be changed by the customer.

In order to draw special attention to these changes, a form, usually of a distinctive colour, is prepared, giving full details of the amendments required. This is issued for attachment to the works instruction ticket, with copies to other departments as necessary.

SUPPLEMENTARY INSTRUCTIONS

Job No. Customer's name ______________________

Description of job ______________________

Attach signed sample if materials affected

PARTICULARS (as full and explicit as possible):

Date ______________________ Initials ______________________

The above corrections have been noted on works instruction ticket

Date ______________________ Initials ______________________

Figure 2.8: Amendments to order/supplementary instructions

3) Daily docket (general)

This docket is designed for use in smaller businesses. For larger organisations dockets for individual departments are recommended. Dockets for origination, printing and finishing are available, each of a distinctive colour.

The daily docket is filled in by the operative and is used for mainly calculating the actual cost of the job when it is completed, although it can also be used for calculating wages.

Daily docket

Name ______________________ Date ____________ 19____

Job No.	Customer's Name	Description of job	Operation no.	Machine	Output	Ordinary Hrs	Ordinary Min	Overtime Hrs	Overtime Min
						Total hours			
Examined by ______				For office use only	Direct hours / Indirect hours				

Figure 2.9: Daily docket

4) Materials release and cutting slip

Details corresponding with those on the works instruction ticket, and bearing the same job number, are given on this form.

It is sent to the materials store when issues from stock are required. If the paper, board or other substrate has to be purchased specially for the job, prior notice must of course be given, either to the warehouse or to the purchasing department.

5) Stock control records

The cost of materials is generally a substantial part of each printing order, being on average about one-third or more of the total cost. Efficient control of all materials and material costs is therefore extremely important.

Essential for the successful control of stock is the maintenance of a smooth and effective system of recording to ensure the availability of the right materials at the right time and in the right place.

MATERIALS RELEASE AND CUTTING SLIP

Job number ______________

Customer ______________ Date ______________

____

Description of job ______________

____

____

____

____

Customer's order no.	Date	Quantity
Delivery required	Delivery promised	Initials

All copy, layouts, specimens, proofs, etc., to accompany these instructions throughout the works.
On completion enter the job no. on Daily delivery sheet and pass to Cost office with (–) file copies

Paper, Board	Stock reference	Quantity	Requisition to stores	Order received in department: Date	Order received in department: Initials

Cutting instructions

For further instructions, see over

Figure 2.10: Materials release and cutting slip

In addition to accurate details of the cost of materials purchased and used, stock records should provide the following up-to-date information:

a) details of stock held;
b) quantities of materials ordered and expected dates on which they will be received;
c) qualities of stock earmarked for orders in hand;
d) balance of stock available for future orders; and
e) an indication of slow-moving, or non-moving, items of stock.

The total value of stock should be easy to obtain at any time for accounting and insurance purposes. Where stocks are subject to substantial variations it is important that insurance cover should always be adequate.

All stores should be in the care of one person responsible for their safe custody and for the recording of receipts and issues as though they were actual cash. Layout of storage areas should provide the best physical conditions and the maximum efficiency in handling. All unnecessary handling of stocks and work-in-progress adds a burden to costs.

Minimum stocks

Although stock control demands the provision of an adequate supply of materials at all times, it is also essential to keep stocks low. Stocks and work-in-progress represent money and therefore a rise in stock must be recognised as tying up cash, which may lead to difficulty in paying creditors and wages. If stocks are not kept at a minimum consistent with production requirements more cash will be required.

Stocks may increase for any of the following reasons:

a) low rate of output in the factory;
b) over-buying of materials;
c) shortage of orders;
d) large orders nearing completion; or
e) expansion of business.

The availability of work, coupled with the rate of production is a vital aspect of profitability. The length of time from the start to the finish of each job decides the volume of production obtainable from the capital employed; thus it decides the profit return on capital. If overall production time can be reduced, capital tied up in materials (and in many other items) will also be reduced. The aim with materials should be to move them through the factory in the shortest possible time.

Costing system records

The BPIF system of cost accountancy provides effective stock control by means of the forms illustrated - form 7, *Daily return of paper and board issued*, and form 8, *Stock record*.

Materials used on each job are recorded directly onto cost sheets; it is essential therefore to have an accurate record of the value of materials used to ensure correct charging.

A stock record (form 8), should be kept for each line of stock held. As fresh stock is ordered an entry to this effect is made and when the material is received and checked, the quantity is added to the balance of stock. When issues are made to jobs, entries from the daily return of paper and board issued are made on the stock record showing the job number, the quantity issued, and the adjusted balance in stock. The quantities on the daily return of paper and board issued are extended at the appropriate cost prices as shown on the stock record and posted to job cost sheets.

Periodically, the balances shown on the stock records should be checked against the physical stock held and any wide variations investigated: stock records may be adjusted for minor differences.

Daily return of paper and board issued

Date				For office use			
Job no.	**Description of material**	Weight 1000s kg	No. of sheets issued	Total weight kg	Price per 1000 £	Total value £	p
			Daily total				

Figure 2.11: Daily return of paper and board issued

The normal working levels of stock entered at the head of the stock record enable close control of stock to be exercised, provided these levels are carefully determined. The figures set should be constantly reviewed to take account of any change in circumstances.

Minimum stock is the quantity below which stock should not be allowed to fall. It should be based upon the number of weeks' consumption that it is considered should be held as a safety margin.

Maximum stock is the level beyond which stock should not be carried and will be governed by financial, market and supply conditions as well as changes in demand.

Quantity to order is that unit of quantity which is most economical or convenient to order.

Re-ordering level is the point in the reduction of a stock where action to replenish should be taken. It is based on the minimum stock plus the estimated consumption during the period which will elapse between ordering and receipt of goods.

Stock record sheet

Description______________________________

____

______ Size _____ mm x__________ mm

Weight______kg per 1000___________g/m²

Price ____ per kilo Net price _____ per 1000

Minimum stock __________ sheets

Maximum stock __________ sheets

Quantity to order__________ sheets

Re-ordering level ________ sheets

Supplier______________________

Stock no. _______________

Stored:

Room _______________

Rack _______________

Shelf _______________

	Ordered			**Issues received**				**Balance stock**		**Issues received**				**Balance stock**		**Notes**	
	Date	1,000s	100s	Date	Job No.	1,000s	100s	1,000s	100s	Date	Job No.	1,000s	100s	1,000s	100s		
	RECEIVED																

Figure 2.12: Stock record

Stocktaking

The dislocation so often associated with the annual stocktaking may be avoided by use of the *perpetual inventory* method. In this system each item of stock is checked physically and compared with the stock record several times throughout the year. If a few items of stock are checked daily, or whenever issues are made, so that every item is checked at least three to four times a year, the records will be sufficiently accurate to provide figures for the annual stocktaking without the need for a physical check at that time. Where an annual physical check cannot be avoided it can be facilitated by the use of a tape recorder into which details and quantities can be spoken. Typewritten lists can subsequently be made from the recorded details.

6) Order form

This is the printer's own order form used when placing orders for outside trade services, such as platemaking and print finishing etc. It is usually prepared in triplicate, one copy being sent to the supplier, one to the accounts department, and one being retained by the originator for reference.

Quality Printers Ltd.
Imprint Works, Newtown
London Office: 1 Caxton Buildings WC1

Order number   to be quoted on all invoices and correspondence   ________19

ORDER FORM

To ____

____

Please supply to the following specifications:

Delivery ____

Price ____

Yours faithfully,
QUALITY PRINTERS LTD.

Figure 2.13: Order form

7) Spoilage ticket

A form is usually employed if work is spoiled through error, accident or customer rejection, so that a report can be made and permission obtained for the additional work to be put in hand immediately to make up the quantity of material or printed matter spoiled.

All spoilage of paper and board, either plain or printed, or materials used in the finishing processes, has to be reported to the supervisor or overseer of the department or section concerned. This spoilage may be the result of an accident, such as spilling tea over a pile of printed work, or of negligence, such as paper having been wrongly cut or wrongly backed-up in printing. The arrangements for issuing additional paper or other materials for reprinting must be carefully controlled and the apportioning of costs to the appropriate departments or house account.

SPOILAGE TICKET Job no. ________

Main work ticket no. ________ Customer ________

Date of spoilage ________ Department ________ Quantity ________

Complaint by ________ Spoilage ticket raised by ________ Date ________

Machine/Operation no. ________ Manned by ________ Shift ________

Reason for spoilage ________

EXTRA WORK REQUIRED AS RESULT OF SPOILAGE Quantity ________

Description ________

State of job at time of issuing ticket ________

Works/production manager's signature

Departments involved ________

DELIVERY DATES: Original ________ Revised ________

Paper/Board requirements ________

Dept.	Period due	Op. no.	Operation description	Overseer's sig. In	Overseer's sig. Out

Operative(s) responsible	Overall cost
	Labour: Materials: TOTAL:

Figure 2.14: Spoilage ticket

8) Despatch advice

A copy of the advice of despatch is sent to the customer, also to the accounts department, so that the total quantity delivered can be checked and appropriately invoiced in due course.

Quality Printers Ltd.
Imprint Works, Newtown
London Office: 1 Caxton Buildings WC1

To ____________ Date ____________________

Your Order no. ____________________

Our Order no. ____________________

DESPATCH ADVICE

We have today despatched the following goods:

Quantity	Description	Per	Consigned to	No. of parcels	Packed in

CLAIMS Your attention is particularly drawn to the notice printed on the back of this despatch advice as to claims against British Rail or the carriers in the event of damage to, shortage of, or non-delivery of goods.

Figure 2.15 Despatch advice

Organisation charts

Organisation charts are a means of illustrating the management structure of an organisation in relation to the responsibilities, authorities and relationships of the personnel concerned.

The organisation chart, being structured to reflect specific requirements, varies in its form to take account of these individual requirements.

Figures 2.16, 2.17 and *2.18* represent three differing organisation charts down to the level of line managers/overseers, for a small- and medium-sized general/jobbing printer plus a large-sized printer.

Managing Director
Sales Manager
Production Manager
Office Manager
Pre-press Overseer
Machine Printing Overseer
Print Finishing Overseer

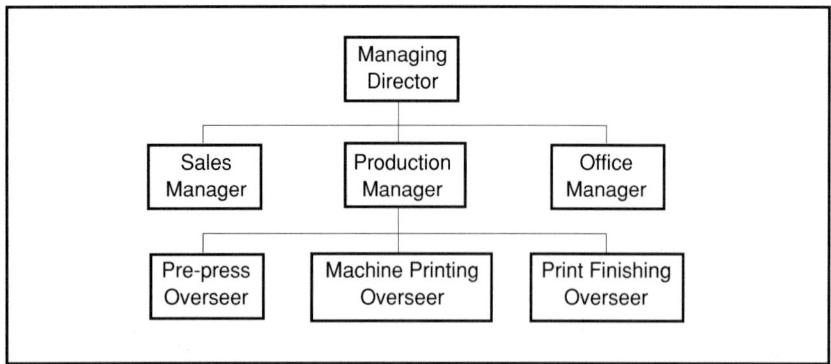

Figure 2.16: Organisation chart suited to a small-sized general/jobbing printing company

Board of Directors
Managing Director
Quality Manager
Administration Manager
Sales Director
Production Director
Customer Support Services Manager
Financial Manager
Sales Manager
Account Executives
Estimating
Purchasing
Order Controller
Costing
Pricing & Invoicing
Production Co-ordinator
Pre-press Overseer
Machine Printing Overseer
Print Finishing Overseer

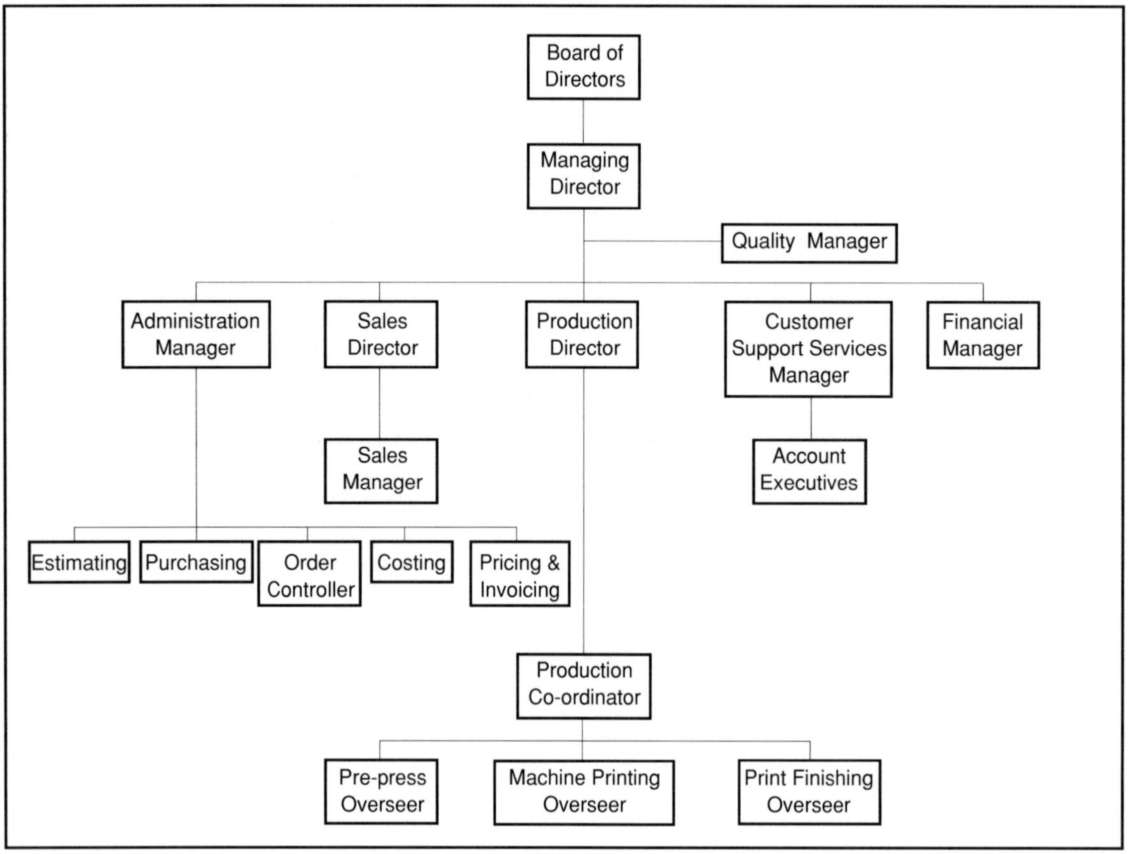

Figure 2.17: Organisation chart suited to a medium-sized general/jobbing printing company

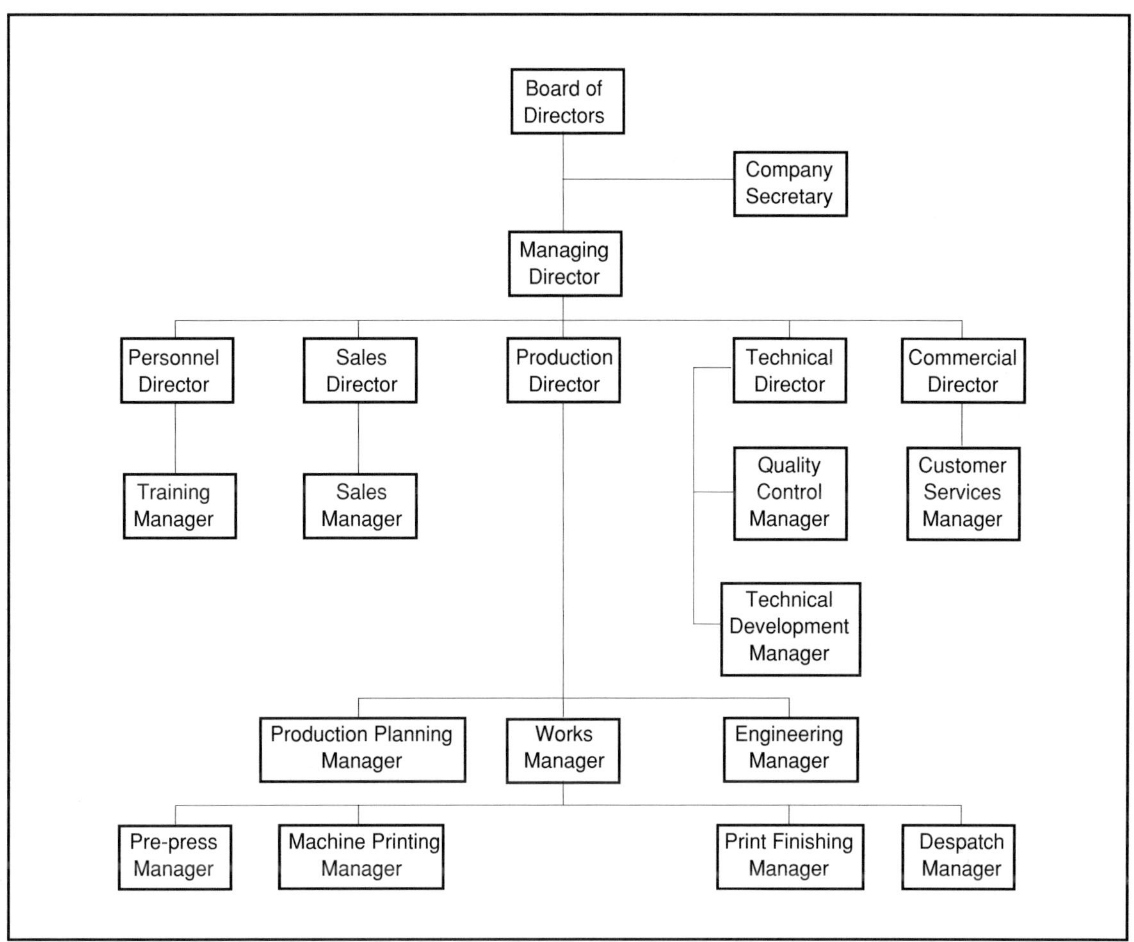

Figure 2.18: Organisation chart suited to a large-sized printing company

3 Management Information Systems

All organisations in day-to-day transactions and communications are faced with a bombardment of information from a variety of sources and angles which, if not carefully harnessed and controlled, will certainly work against plans for continued success and future growth.

Well into the 1990s, it is not unusual for small and large organisations alike to suffer from 'information overload'. Manual and non-integrated systems cannot cope effectively with this information explosion, whereas, when installed and fully operational, a MIS *(Management Information System)* provides a framework and set of tools to receive, store and analyse information which responds to users in a constructive and meaningful way.

An MIS can supply, for example, a wide ranging and comprehensive database of invaluable customer information. Every working day, on an on-going basis, a tremendous amount of information becomes available on potential, new, long-standing and 'lost' customers.

Information and communications are flowing freely each day, and often through the night in the form of faxes and electronic mail, between different personnel in the long chain of business communications - sales, production, administration, accounts, customers, print buyers, specifiers etc. - all recording their needs in verbal, written or some means of electronic link.

For organisations without an MIS this remains a largely untapped source of information and feedback on which current and future strategies can be productively based. This chapter covers a brief outline of what an MIS is and what advantages it should bring to an organisation by considering the following specific areas: MIS; types of MIS available; benefits derived from operating an MIS; impact on personnel and the organisation, and Electronic Data Interchange.

An MIS (*Management Information System)* consists of a network of computers linked to an organisation's host central database unit used for recording, accessing, storing, manipulating and processing information, so allowing all levels of managers and users the opportunity to perform more effectively.

It could also be defined as a *computerised, integrated system* which provides *management information* and is capable of performing a *range of functional administrative tasks* - the result being an extremely effective management tool which is outlined in greater detail overleaf.

Computerised

Computers have the facility to store vast amounts of information in 'structured' form, calculate at incredibly fast speeds and present data in a wide variety of forms which result in tremendous benefits to managers at all levels - an MIS incorporates all of these facilities.

Integrated

The MIS, when at its most effective, contains elements that interact and automatically update related data within the system, so alleviating the need for multiple alterations and changes - eg hourly cost rate changes only needing to be done once, with the result that estimating, costing and accounts departments can be sure they are working to the same rates at any given time.

Management information

Management information which is accurate, accessible and concise, forms the lifeblood of any organisation: it therefore needs to be sufficiently extensive and relevant to allow short-term operational and long-term planning decisions to be relied on - a well maintained MIS meets these requirements.

Range of functional tasks

When an MIS has been installed it should be capable of undertaking a wide range of functional tasks including those most likely to be required by the organisation concerned.

To gain the maximum benefit of an MIS, information needs to be recorded, compared, analysed and updated in *'real time'* - ie processing information, through all the required stages, immediately it is available and/or as the event happens or is completed. The most effective means of achieving *real-time* recording is by introducing electronic links such as keypads, direct machine monitoring and bar code reading - for further information see *Chapter 6, Production control,* 'Monitoring of production data', pages 100 and 101.

Types of MIS available

Computerised management information systems fall into two types:

a) off-the-shelf or standard

b) customised or bespoke.

There is, however, a considerable amount of middle ground where the two systems merge, especially when an organisation updates and customises its existing basic system by adding further modules or programs which are in some way 'customised' to match individual requirements. The range of systems available to printers is increasing rapidly, with the choice of system to be installed by an organisation depending largely on the size, scope and type of MIS they wish to operate.

a) **Off-the-shelf systems** involve a range of standard packages or modules which allow only minor customising in the form of hourly cost rates, equipment range, stock sizes and prices etc.

The most popular 'off-the-shelf' systems are *modular*, which allow each printing company to enter at whatever level and cost appropriate to their current circumstances. Being modular in construction the MIS installed would normally be capable of growing with the company, additional users and modules being added as and when required.

The starting-point or entry-level for smaller-sized printing organisations is frequently a one or two PC-based system with estimating, costing and possibly works instruction documentation modules. For the larger-sized printer a six to eight PC-based system is common, with the additional modules covering sales order processing, works instruction documentation, production control and stock control.

Often a system is chosen which will link with existing accounts packages; alternatively, accounts and payroll, etc will be run separately until a fully integrated MIS is considered.

Some of the more established suppliers of Management Information Systems have produced a suite of programs which have been 'bespoked' to suit an organisation's particular requirements - eg separate estimating modules for sheet-fed and web-fed offset-litho printing, reel-fed labels, cartons, screen printing and pre-press requirements only etc.

b) **Bespoke systems** are written specifically for individual organisations. The bespoke system is normally developed to follow the pattern or sequence of the organisation's previous manual system. This can often make the acceptance of the system into the organisation easier as users can adapt quicker utilising their previous knowledge and skills to the new integrated computer-centred system.

However, the major drawbacks associated with a bespoke system, compared to an off-the-shelf system, is that it is more expensive and involves a lot of development time.

Few systems introduced into printing companies are truly 'bespoke' as the suppliers of larger systems adopt the strategy and approach that they have developed through working with similarly based organisations.

Most larger computerised systems are built upon a model of the organisation, where the user can 'pick and mix' the building blocks required to cover the processes, machines, materials, labour and financial systems used.

Software and hardware can often be upgraded in a modular path with many suppliers moving towards an open architecture platform such as UNIX, also fourth generation languages which take the form of a set of tools for writing programs that avoid much of the detailed instructions and programming associated with previous languages such as COBOL.

Benefits derived from operating a MIS

The main areas that can be identified as gaining the maximum benefit from introducing an MIS, as against operating a manual non-integrated system, are *sales and marketing; administration; production planning and control linked to productive efficiency and workflow; financial and cost control.* The benefits are highlighted in italics, followed by a short commentary.

Sales and marketing

Identification of markets which produce the best results for the organisation
- indicated by an analysis of previous sales-related-data broken down into different job types and market sectors as required

Assistance with the planning of sales campaigns by month, product, industry sector and geographical area
- again allows targeting based on historical data built up in the system

Ability to predict several possible prices using different criteria
- such as value added, target mark-up and projected market rate

Improvement in the success rate of estimates
- due to quicker turn-round and ease of access to previous conversion/ success rate plus recording of competitors' rates when available

Improved measurement and control of sales, including marketing processes
- provision of a wide range of data built up from sales prospects, market research and penetration used to direct and monitor marketing and sales mix.

Administration

Improved internal communications through the use of a centralised store of information
- increased interaction and feedback as all users work from the same database

Tighter control due to data validation and standardisation
- individuals and departments take much greater care in ensuring the data entered is accurate and to approved specification as everyone relies on its accuracy

Enhanced customer service through the wider access and availability of information
- much wider ranging and higher quality customer information available from historical data

Improved documentation through consistent presentation of information
- pre-structured set-up of all data, so ensuring all users work from the same databank

Support of quality management systems through implementation of disciplined procedures and audit trails
- the 'traceability' element is improved with the use of an MIS

Long-term time savings through improved efficiency
- benefits are considerable from when the system is first installed, and increases as more and more data becomes available.

Production planning and control/productive efficiency and workflow

Optimisation of production resources through improved scheduling of available work
- use of a production scheduling module on an MIS normally allows 'what if' scenarios, also lists-of-jobs can be produced indicating the most effective ways of working across a wide range of machines

Improved control of materials, purchasing and usage
- through supporting documentation and system verification, as follows:
- recording of stock receipts, issues, usage and returns, especially using electronic means of stock movement recording such as bar code reading
- ensures highly efficient stock control and monitoring

Earlier identification of potential bottlenecks due to improved shopfloor information
- if 'real time' recording is used, production controllers are in a position to respond quickly to discrepancies between estimated and actual times in re-arranging current and future work plans; also forward loading can be 'spread-out' much more realistically with 'real time' reporting

Improved customer service through better prediction and fulfilment of delivery dates
- with better feedback reporting and control over production activities
- especially if 'real time' is in operation - customer service in terms of improved communications, reporting and achieving target delivery dates should be significantly improved.

Financial and cost control

Improved credit control through up-to-date work-in-progress and debtor information
- bringing together all production, materials, administration and financial data into one common system - ie the MIS - ensures accurate information on any matter is quickly established

Improved facilities for the collection and analysis of costing and charging information
- replacement of recording manually-operated time sheets/daily dockets by electronic means such as keypads and terminals. This improves the accuracy, speed of data entry and reconciliation associated with costing and pricing policies

Potential for speedier billing, thus improving cash flow
- long delays associated with lengthy end-of-month charging procedures are considerably reduced as invoicing can take place at any time on an individual job once it is completed and all the associated costs are booked to it

Elimination of multiple input of financial transactions (where links to financial ledgers are available)
- an MIS should ideally contain the facilities to complete all the necessary financial operations required by a printing company within its integrated financial/accounts module or link into a compatible package such as Sage, Multisoft or Pegasus

Improved materials/outwork purchase and financial control
- more accurate job requirement forecasting and recording of stock movement helps keep purchases closer to minimum stock/outwork holdings or JIT (*just-in-time*) targets.

Figure 3.1 illustrates a flow diagram of a comprehensive MIS which would be appropriate to a large-sized company or organisation operating a fully-integrated system.

Impact of an MIS on personnel and the organisation

Different departments and functional areas in an organisation, especially any area involved directly in customer service, such as account executives, can work in a much more effective and informed way through the use of an MIS, so improving the quality of response and service given to the customer.

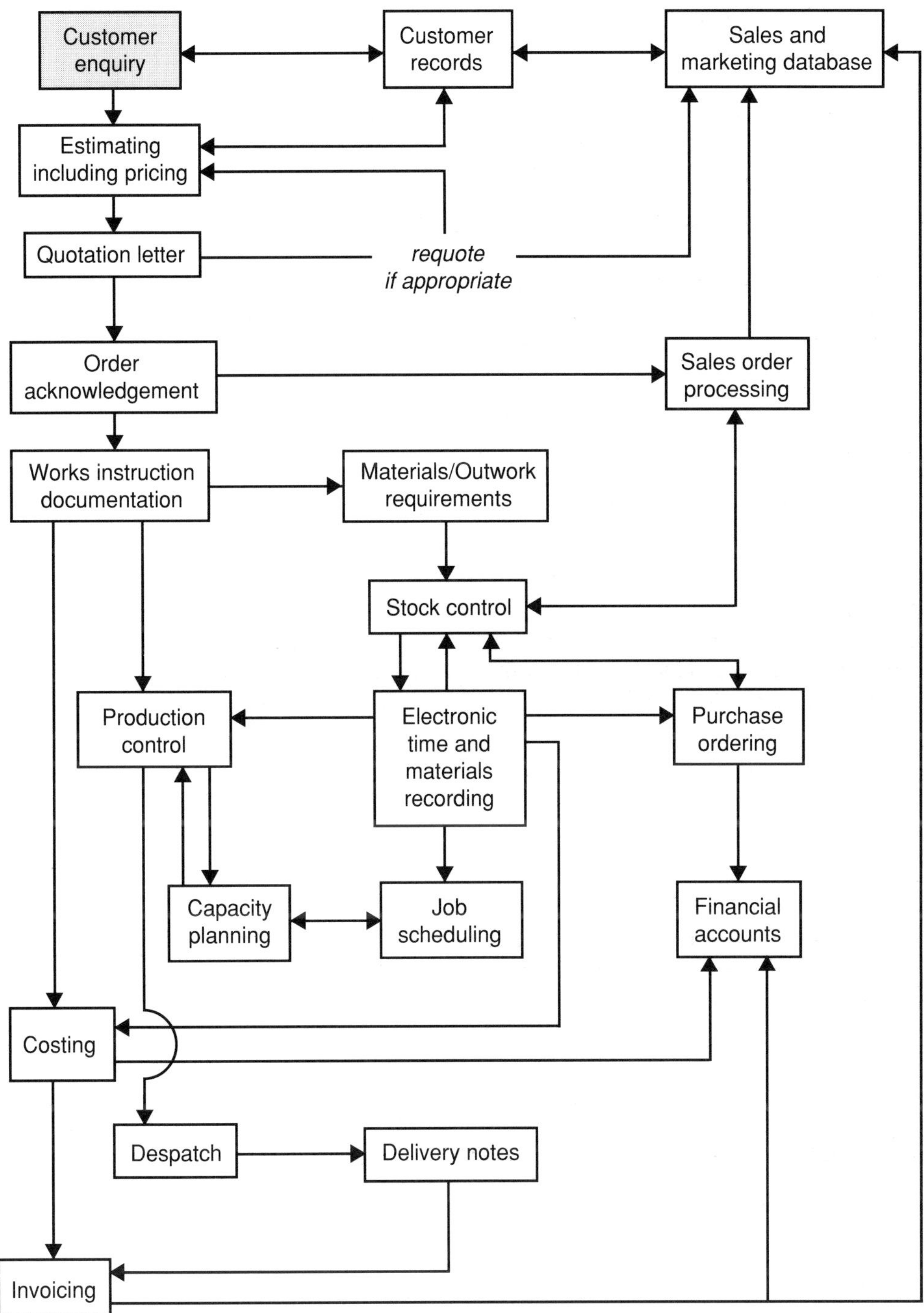

Figure 3.1: Modular configuration of a comprehensive Management Information System

With a fully functioning system, most MIS users would admit that it is a case of the whole being greater than the sum of the parts, ie the collective benefit to everyone is greater than the individual inputs of each user.

The information available to individual departments is invaluable in completing their specific role effectively, but the cross-directional data also proves to be beneficial in letting all departments and users see how their role contributes and interacts with each other.

One of the approaches of BS 5750/ISO 9000 is to encourage the aspect of '*internal customers*' where every individual is seen as a link in the *quality chain* and is responsible for ensuring that the quality of their product or service is to the required standard and specification before being passed on to the next part of the chain. This self-supporting aspect of individuals and departments alike is very much central to the philosophy of a management information system.

It therefore needs to be acknowledged that for the MIS to operate to its full potential it requires the full co-operation and understanding of all employees and users. Individuals must ensure their recording and inputting of data is correct and accurate as the system is only as good as the accuracy of the data contained within it.

Figure 3.1 highlights the interdependence of departments and individuals in fulfilling the *complete print chain from enquiry to invoicing, including the important areas of feedback.*

EDI

A further development in the sharing and accessing of data for control and information purposes is the use of EDI *(Electronic Data Interchange)* which is the electronic transfer of data between computer systems in different companies and organisations.

Such facilities are bound to grow as suppliers and customers seek faster and more accurate ways of transmitting and receiving information from each other.

Basically, EDI is an electronic link, such as a modem, between parties which, for example, would allow customers and the printer to send and receive orders and instructions, or for the printer to link up with a paper merchant's stock holding. BACS *(Bank Automated Clearing System)* is a form of EDI. For further information on EDI, see *Chapter 13, Sourcing suppliers and purchasing.*

4 The role of the account executive

Just as changes have been introduced to streamline and improve the efficiency and productivity on the 'shopfloor'/production areas, so the process of change has had a major impact in the areas of administration and management of printing companies.

There has been a growing awareness for some time among many printers that the quality of service they offered needed to be improved, so that members of the administrative and management teams were fully aware not only of their own role and responsibility, but also how it fitted into and co-ordinated with other individuals and departments.

The *Introduction* raised the point of the 'specialist' roles of estimators, order controllers, production planners, costing and charging co-ordinators still being retained by the medium- to large-sized printing companies but at the same time recognising that more and more companies are introducing the role of the *account executive* as a focal point of contact for customers. In the larger-sized company the account executive will normally act as the *co-ordinator* of all the functions related to a job from receipt of an enquiry through delivery to the customer and after-sales follow-up. The role of account executives within small-to medium-sized printing companies differs somewhat in that they are required, in addition to their central role of 'inside' customer contact and service, to perform many of the 'specialist' tasks such as estimating, raising works order documentation, progress chasing etc.

The introduction of an MIS *(Management Information System)* - see *Chapter 3* - has helped speed up the process of the account executive taking over this relatively new and developing role. Even where only mainly manual administrative systems are operated, the central role of the account executive is still seen as having a major part to play. In this chapter the term account executive will cover this central administrative role; however, many job titles are used in this capacity in the printing industry, such as production co-ordinator, customer contact, customer support or services representative and account controller - the reader is therefore advised to adopt whichever title they feel most comfortable with. Whenever an overall departmental role is implied the term *Customer Support Services* will be used.

The topics covered in the chapter are as follows: communication and interaction in print order processing - interaction and communication with customers, representatives, 'in-company' printing departments and outwork suppliers; skills and qualities required by the account executive - estimating, costing and pricing policy.

The particular part which the account executive has to play in the structure of a printing organisation can be seen from *Figure 4.1*. This is at the centre of both administration and the production departments liaising with customers, sales executives, production departments, outwork suppliers and services.

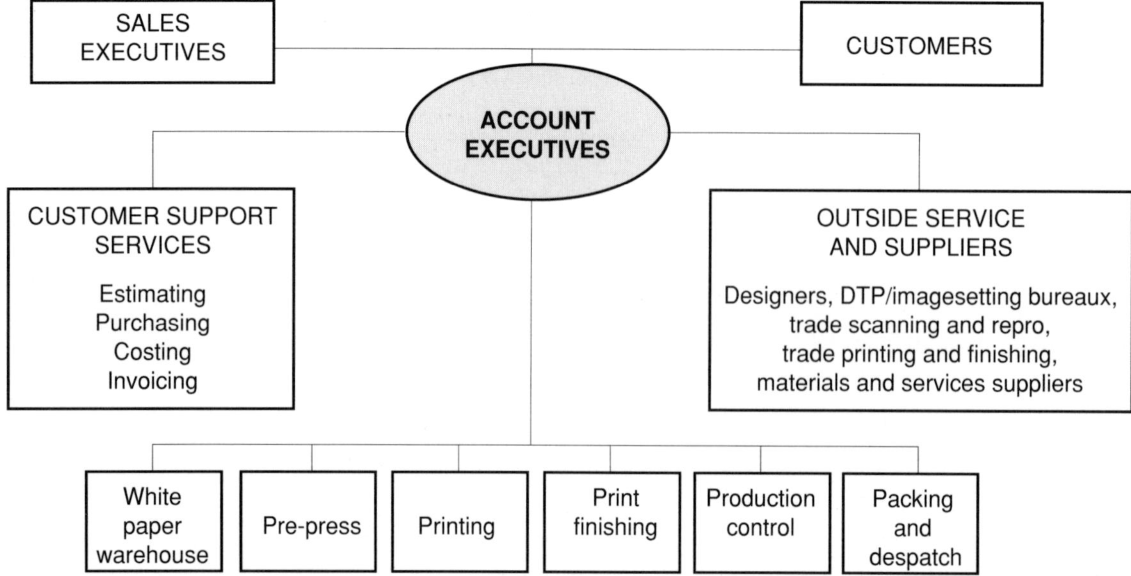

Figure 4.1: The account executive as the central point of communication

Communication and interaction in print order processing

Criticisms frequently raised about printing companies by customers include long delivery dates and broken delivery promises; however, the present-day printing industry, which has embraced new technology in such a big way, uses a much more predictable series of processes.

With tighter control and a more professional approach, improved customer satisfaction through shorter lead times and better forecasting should be possible with the help and co-operation of all parties involved in the printing production chain. Being mainly a 'bespoke' - ie made to special order industry, each item of print produced presents a fresh challenge.

Ambiguous instructions in any part of the communication or interaction chain, from the inception of an idea to a finished printed product, can lead to costly mistakes, frustration among all parties concerned and lost customers. Correct, efficient and orderly instructions help to prevent or at least minimise jobs going wrong and ensure that they are produced to everyone's satisfaction. Interaction implies reciprocal action or feedback between parties to achieve a given purpose. The chain of communications in producing printed products can be protracted and it is therefore crucial to control and record the process at every stage.

The start of the print communication chain is the customer or print buyer who, more often than not, passes print enquiries or orders through the printing company's sales executive or representative. From there they may pass to the estimator or works order department. If proceeding as a print order, the production control function will track the job through all the required departments, including those for materials and outwork provisioning - see *Figures 4.2, 4.3* and *4.4*. At all stages *feedback* between some or all of the parties concerned must be maintained.

Interaction and communication with.....

..... *Customers*

Without customers there would be no business. Every customer must be considered not only for their current order but as potential for future business. Getting it right each time will secure the next order.

Some customers will only order print very infrequently - eg a catalogue published once a year or batches of stationery every few months. At the other extreme is the professional print buyer or publisher who has an annual print-buying budget of hundreds of thousands or millions of pounds. Infrequent buyers of print will rely heavily on the printer's staff to guide them through the stages of printing and the onus is very much on the account executive to maintain close contact and provide reliable advice, interaction and communication at all stages.

In the case of experienced or professional print buyers, the printer is dealing with customers who work with and through a wide spectrum of printing companies: they will certainly lay down well-documented instructions and procedures, expecting in return a printer to understand them and to respond positively to a well-structured relationship.

..... *Sales executives*

The sales executive's main function is to make and maintain contact with customers and to create a demand for a printing company's products and services by attending to customers' requirements. The customers serviced by sales executives will range from very long-established accounts to potential new business.

To be successful in handling customers, sales executives must be aware of their limitations when it comes to matters of a complex technical nature. Sales executives are the *bridge or external line of communication* with the customer and it is essential that the interaction with the printing works is efficient. The sales executive must ensure the correct interpretation on each enquiry or order is passed onto the account executive, preferably by the completion of an enquiry/order form, with all the essential information supplied in the required format.

..... *'In company' printing departments*

'In company' printing departments fall into two main groups - *office or non-production* and *works or production departments.*

The main office or administrative function areas cover estimating, works order control, production planning and control, purchasing, costing, pricing and invoicing. All these functions form the direct continuum of the progress of a printed job. Each area, possibly apart from costing, will have need to liaise with the customer direct or through the account executive and will therefore have a contribution to make towards efficient print order processing.

The works or production departments include the areas of pre-press, machine printing and print finishing. The production planning and control department acts as the *interpreter and controller* between the main office areas and the works.

In an attempt to streamline and improve communications with customers there is a growing trend for printing companies to appoint account executives to look after a particular group of customers. The account executive is the *inside works contact for the customer* and it is his or her responsibility to progress and service each enquiry and order for the customer. There is considerable merit in this arrangement because account executives and sales executives become thoroughly familiar with the particular requirements of each customer. Account executives who have been trained as estimators or order clerks can also carry out these duties when required. Customers find it of considerable benefit when contacting the printing company to be able to ask by name for the person who handles their work throughout the complete job cycle.

..... *Outwork/materials suppliers*

Outwork and materials suppliers, although outside the structure of the printing company, form a central and invaluable source of goods and services, without which the full print service could not be completed.

Departments within a printing company have the advantage of constant formal and informal communications with easy access and face-to-face contact when required. This is not possible with outside suppliers, and it is therefore essential that accurate provisioning, scheduling and timetabling feedback is maintained at all times.

Any change to instructions or projected delivery times must be passed on to the other party as soon as this information is available. Any gaps in the communication process leads to customers being misinformed, let down and ultimately becoming lost to the company.

Knowledge, skills and attitude required by the account executive

As the account executive is the link between the customer's requirements and the company's functions, he/she requires particular attributes to be successful. The following is a typical list of those attributes:

Knowledge

- knows what is possible in terms of design and production of printed matter

- has a clear understanding of the relationship between print and various substrates

- understands cost implications of changes to specified work

- knows his/her own company's production procedures and requirements

- understands the legal requirements regarding printed matter.

Skills

- listens accurately to the customers needs

- is astute at questioning to gain clear instructions

- can interpret production requirements to assist customers check and amend their specifications

- can manipulate data and numerical relationships quickly and accurately

- can influence internal functions to get the job done

- can cope with constantly changing priorities and a large number of jobs simultaneously

- communicates effectively by telephone, face-to-face and in writing.

Attitudes

- believes that the customer comes first, and is sensitive to their needs

- cares for customers, and believes every order well done will lead to another one

- recognises that when the customer calls, the account executive is the company.

Summary

The account executive has a clear role in influencing the organisation so that it can meet customer needs, profitably.

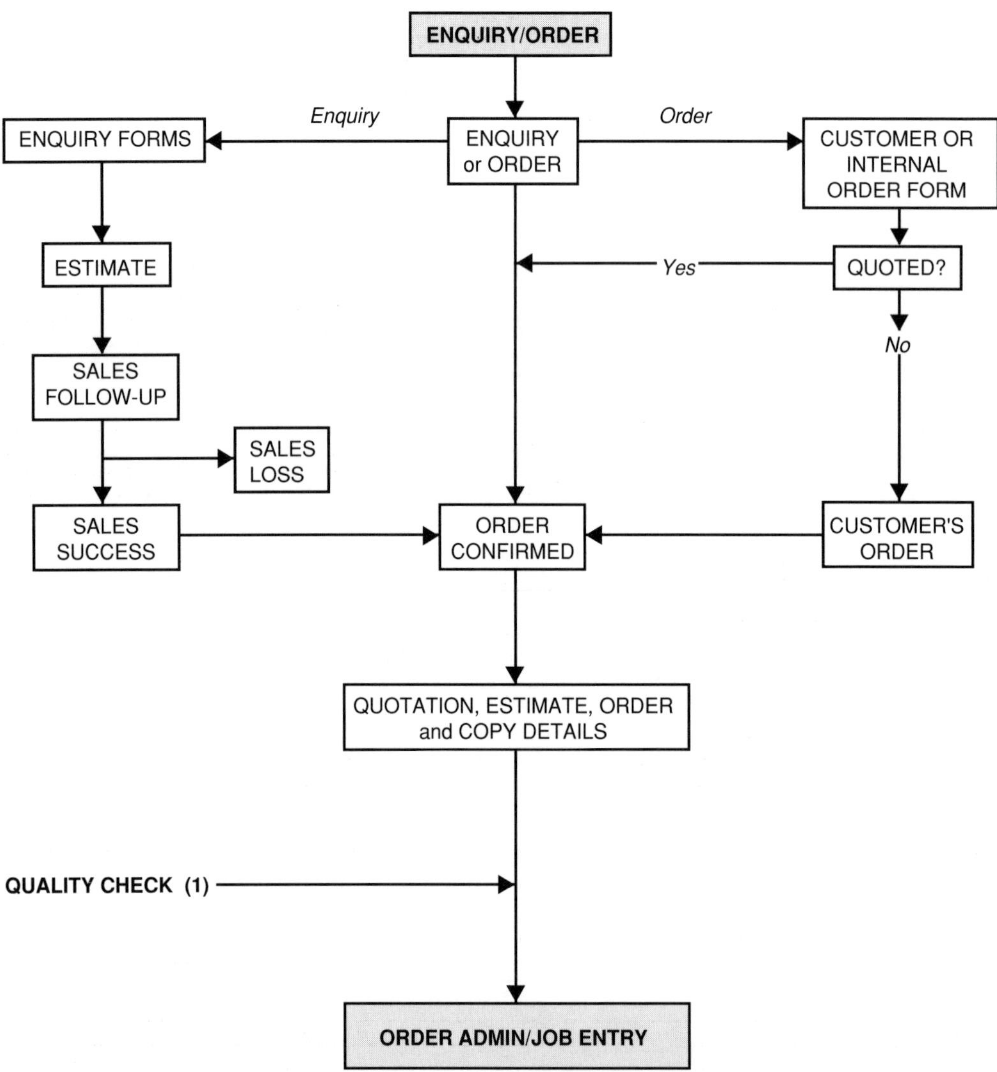

Figure 4.2: Flow-chart illustrating enquiry/order to order admin/job entry

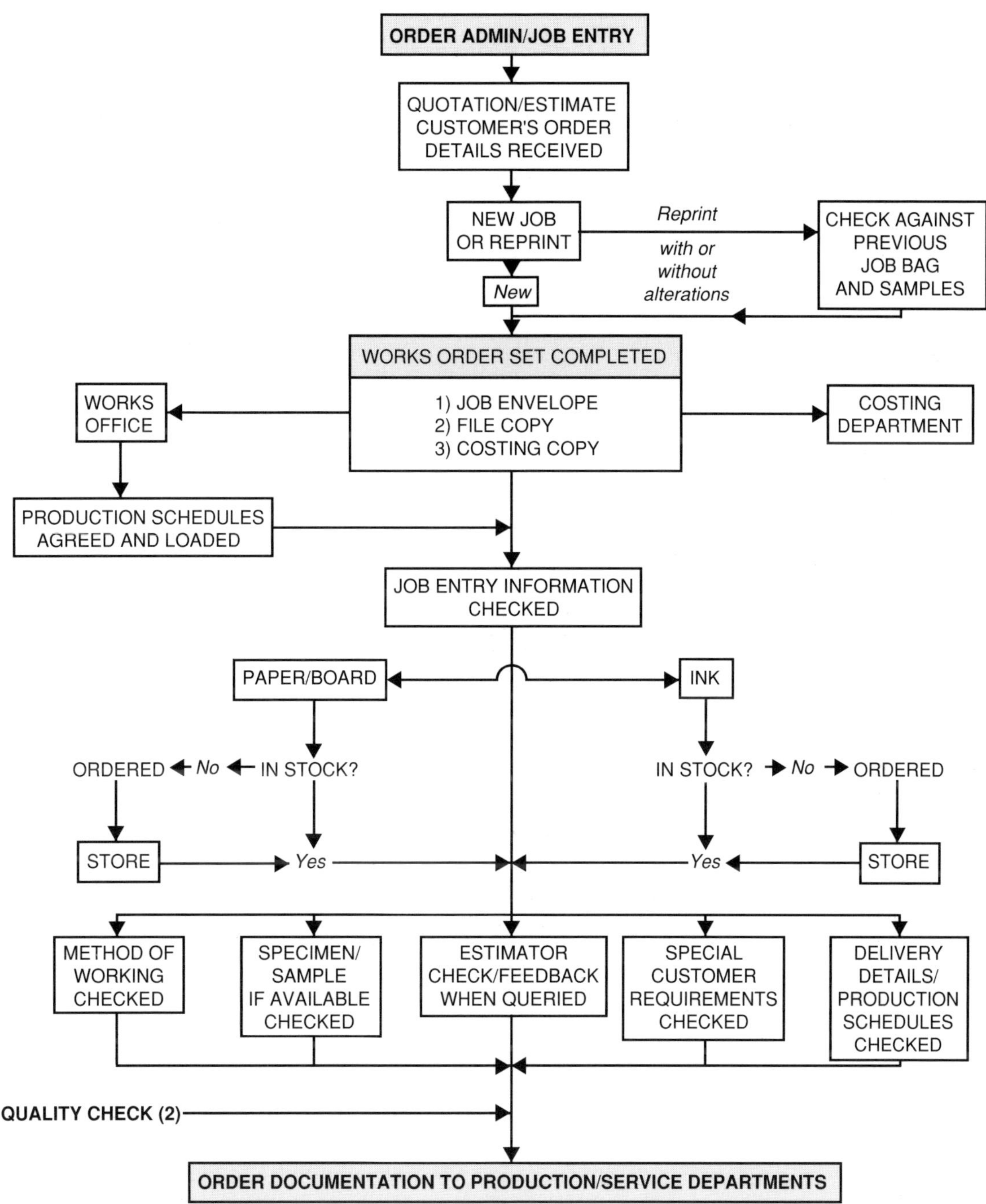

Figure 4.3: Flow-chart illustrating order admin/job entry to order documentation released to production/service departments

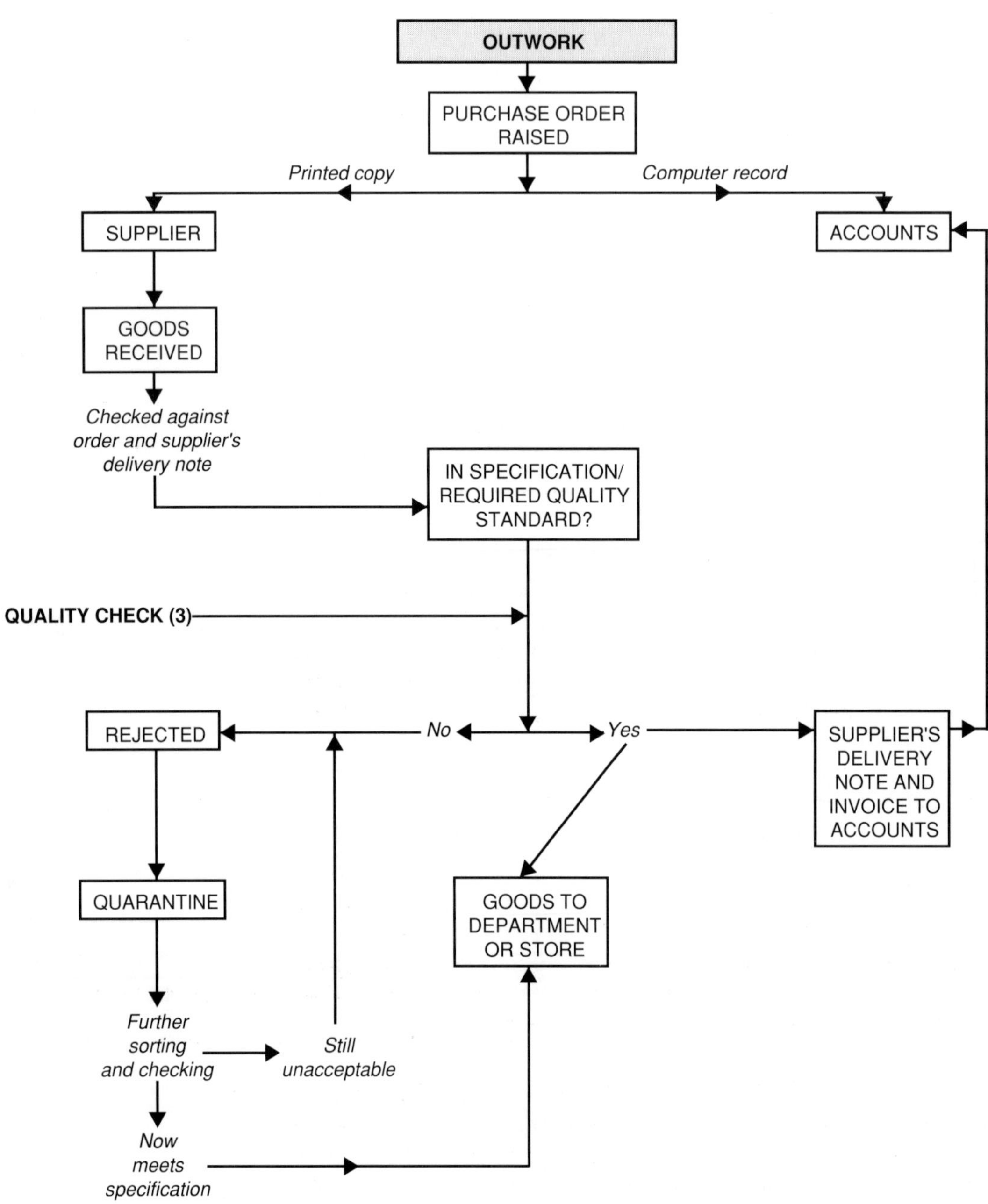

Figure 4.4: Flow-chart illustrating outwork purchasing and processing procedures

Estimating, costing and pricing policy

Estimating, costing and pricing policy are commercial activities of which account executives should have at least an awareness and, at best, a working knowledge.

Estimating

In recent years the 'mystique' of estimating has been considerably reduced by the adoption of computer-assisted estimating systems often linked to a company's MIS.

Such systems rely on creating a database built around company specific production and financial criteria such as production machinery - size, speed ranges, number of units etc; manual operations and procedures such as film planning, retouching, platemaking and benchwork; material costs and usage; products and services; hourly cost rates, mark-up percentages - variable and fixed, depending on company policy.

Once the company specific (bespoke) data is input into the system, the host software arranges it into a form which is relatively easily accessible to the user. Prompts, guidance and help screens, often menu-driven, guide the user through the process of creating an estimate. There are normally three main options for creating an estimate on a system - either from *scratch* - ie completely new; *adapting existing job data* closest to the current enquiry; or from a *series of templates* which often take the form of a range of 'model' estimates representing typical jobs undertaken by a company. These can then be altered, similar to adapting existing job data, to suit the requirements of new enquiries.

Some estimating programs have been developed which can be run on 'lap top', small, portable but nevertheless powerful computers, and can be used in any remote locality as required, even on the customer's premises. Sales executives and/or account executives find this another aid to improved customer service and a help in closing the sale.

The comments made earlier about estimating should not be misunderstood in that skilled 'specialist' estimators are still required in most printing companies due to the complexity of the work undertaken throughout the industry. Their expertise will be required to help set up and maintain any computerised system, but a flexible and open system will allow others, such as account executives, to prepare estimates, if only in a limited way such as amending existing quotations which are already on the system.

A major point to be considered with the use of a computerised estimating system is that *it is a tool to assist the estimating function; it does not replace it.*

There may be a degree of de-skilling in the basic elements of preparing calculations but, overall, computerisation provides a skilled estimator with the opportunity to learn and apply new skills. It is the combination of the competent estimator and the computerised estimating system which enhances and extends the estimating function to a higher level than could ever be achieved by just a manual system.

Costing and pricing policy

Because of its essentially bespoke nature, printing has traditionally been a *cost plus/production driven industry*. Due to increased product specialisation, the focus has moved at least partly towards a market-driven approach.

This changes a printing company's approach from being predominantly a passive order taker with fairly standard profit mark-ups, to one which identifies printing product areas where their particular range of equipment and expertise leads to *market sensitive pricing* at a level where higher contributions and profits along with more sustained work flow can be achieved.

If a company adopts a market-driven approach it will identify and become more sensitive and aware of opportunities and threats inherent in pursuing a policy of obtaining any type of work as long as it keeps the presses running. This policy often leads to low prices, costing losses and unbalanced uses of resources, also expensive overtime and/ or expensive outwork charges.

To be successful and profitable, a company needs to identify as much, if not more, with their customer's market base and product area, as with their own. If a printing company only sees itself as putting ink on paper no matter what the potential of the finished printed item, then a probable opportunity of higher profitability is being missed.

Adopting market pricing will result in different profit margins, depending on the specific printed product - eg labels, magazines, cartons, continuous stationery etc, with the result that identifying areas of growth and the benefits of specialisation will improve the company's profitability. A planned mix of work is preferable and more predictable than relying entirely on ad hoc orders.

A review of a company's *cost statement* will show the type of work undertaken previously, broken down into product types. This will reveal surplus on job costs per product type, also sales total and value added. By actively pursuing a policy and a directed approach to customers and product areas which yield the best possible returns, a company can increase its market share in areas which maximise their profitability and use of resources.

If a company adopts entirely the *full absorption costing approach* then cost plus *a percentage profit* often determines the quoted price, especially in the absence of any further information on the job in question. The *estimate* is an opinion or forecast of what the job is likely to cost in terms of labour and materials, with the *price* often fixed by a senior member of management applying company policy in terms of required return etc.

With this approach there is a considerable amount of high and low-pricing. Printers constantly hear about high-pricing with quotations, but very seldom if the price submitted is considered low. Low pricing is an opportunity lost and one which may have an accumulative effect on future prices.

Full absorption costing is often criticised for being relatively inflexible in supposedly recommending a standard mark-up on all jobs and in being unresponsive to market pricing. This is unfair as the mark-up can and should be varied to suit the circumstances of each individual job.

Marginal costing separates out fixed and variable costs, encouraging prices to be seen as making a *contribution* to the company. Once the variable costs have been covered, the balance is set against the fixed costs. Pricing with marginal costing tends to be more flexible, often leading towards specialisation and an 'economies of scale' approach where the higher the volume, the wider the spread of fixed cost recovery. This approach often achieves its objectives by high volume, lower profit margin work.

Value added (sales less cost of direct materials and outwork) is a concept which is gaining popularity in establishing a more cost-effective pricing policy. Value added differentiates between in-company and outwork, so as to identify and highlight the extra value of work done within the company.

Pricing will then reflect this balance when considering if it is viable to take on jobs at a certain price level.

If the overall possible market price for a job works out at £25 000, including £15 000 of repro outwork such as colour scanning, planning, platemaking and proofing, plus £5000 of materials made up mainly of paper and ink, this results in £5000 value added - ie 20% when expressed as a percentage. This is a very low value added ratio, leaving only a relatively small contribution with very little room for manoeuvre in price reduction.

On the other hand, if the overall market price is again £25 000, but this time the repro outwork element is £3000, plus as before, £5000 for materials, this would result in £17 000 [£25 000 - (£3000 + £5000)] value added (68%). This is a much healthier value added ratio which allows greater room in price adjustment if required, thus emphasising the point that value added is a more positive measure of efficient trading than turnover.

For further in-depth coverage of estimating and costing there are two comprehensive BPIF publications available - *Estimating for printers* and *Getting the measure of your business.*

Figures 4.2, 4.3 and 4.4 reproduced on pages 54, 55 and 56 take the form of flow-charts illustrating the possible procedural routes and checks undertaken from the receipt of a customer's order or enquiry to raising the required internal and outwork order documentation.

These cover the main areas account executives will be expected to co-ordinate, in whole or in part.

All three flow-charts are examples of the diagrammatic approach adopted by many printing companies in preparing their quality/ checking procedures.

5 Legal matters

In recent years, printing companies seem to have been bombarded with new or amended laws and legislation of which they have had to take account in ensuring that the working environment and practices within their premises and sphere of influence come within the current laws of the land. The main changes directly affecting employers in recent years have come in the areas of employment, environmental and health and safety legislation.

The purpose of this chapter is to address the main legal considerations and obligations which impinge on print-related matter in its many forms.

The following represents extracts from the BPIF publication *Law for Printers - a guide to legislation affecting the printing industry*; if an unabridged version of the subjects covered, including many other areas is required, reference should be made to this publication. The areas marked with a '●' in this chapter have been reproduced in their entirety.

Copyright

Under the *Copyright, Designs and Patents Act 1988*, copyright subsists automatically, without the need for any formality (such as official registration), in every qualifying original literary, dramatic, musical and artistic work. Briefly, the qualification requirement is met if the author is *either* British *or* resident in the UK when the work was first published or at the time of death if the work was unpublished. The qualification extends to citizens, subjects and residents of other countries specified in Orders-in-Council (including countries adhering to the Universal Copyright Convention). Copyright also applies to all Government and Parliamentary works and to the works of certain international organisations.

Copyright means the sole right to reproduce the work in any material form or to publish it or, in the case of literary work, to make any adaptation of it (such as translating it, or converting a non-dramatic work into a dramatic work).

'Literary works' include any tables or compilations and computer programs. 'Artistic works' include photographs, paintings, drawings, diagrams, maps, charts or plans, engravings, etchings, lithographs, woodcuts or similar works.

A 'photograph' is defined as meaning a recording of light or other radiation on any medium on which an image is produced or from which an image may by any means be produced, but it does not include any part of a moving film.

Copyright is not a right in novelty of ideas. It is based on the right of an author or artist to prevent another person copying an original work which they themself have created. There is nothing in the notion of copyright to prevent a second person producing an identical result (and enjoying a copyright in that work) provided it is arrived at by an independent process. There is no statutory obligation that a copyright work should bear any indication that it is copyright.

Term of copyright

Generally copyright expires 50 years after the year of the author's death, but there are special rules in relation to Crown copyright, works of unknown authorship, computer-generated works and certain works in existence before 1st August 1989 (including unpublished literary, dramatic and musical works, lithographs, prints and similar works and photographs).

Ownership

The author of a work is usually the first owner of any copyright in it. However, except as mentioned below, where a work is made by an employee in the course of his employment, his employer is entitled to any copyright in the work, subject to any agreement to the contrary.

If a work was made by an employee before 1st August 1989 in the course of employment for publication in a newspaper or periodical then, in the absence of any agreement to the contrary, the employee is entitled to copyright except insofar as it relates to publication in a newspaper or periodical. Also, where before 1st August 1989 a person commissioned the taking of a photograph, or the painting or drawing of a portrait, or the making of an engraving or lithograph, he is entitled to any copyright of the work.

The owner of a copyright may assign it, wholly or partially, to someone else, but an assignment is not effective unless it is in writing. Licences, other than exclusive licences, do not have to be in writing.

Typographical arrangements

Copyright also subsists in the typographical arrangement of a published edition. The publisher of the edition is entitled to this copyright, which expires 25 years after the year in which the edition was first published. It is an infringement of this copyright to make any unauthorised facsimile copy of the arrangement.

It should be noted that the publisher's copyright in the typographical arrangement is separate from, and additional to, any other copyright which may subsist (or which may have expired) in the work.

Fair dealing

Fair dealing with a literary, dramatic, musical or artistic work for the purposes of research or private study does not infringe copyright in the work or in the typographical arrangement of an edition. However, multiple-copying is not permitted, that is where the person doing the copying knows or has reason to believe, that it will result in copies of substantially the same materials being provided to more than one person, at substantially the same time, and for substantially the same purposes.

Fair dealing with a work for the purpose of criticism or review is also permitted provided it is accompanied by a sufficient acknowledgement, that is the title or description of the work in question, together with the name of the author.

Typeface designs

It is not an infringement to use a typeface in the ordinary course of typing, composing text, typesetting or printing. Nor is anyone prohibited from possessing the means for reproducing a typeface for such use. This is so, notwithstanding that an article is used which is an infringing copy of the work.

However, it is an infringement of copyright to, without authority, manufacture, import or deal with articles specially designed or adapted for producing material in a particular typeface, or possessing the same for the purposes of dealing with them. This applies for 25 years after such articles have first been marketed.

Moral rights

An author has the right to be identified as the author of a work, and the right to object to any derogatory treatment of his work. The rights are not applicable in relation to works whose author died before 1st August 1989, or acts done in accordance with any assignment or licence granted before that date.

The right to be identified as author applies only if it is asserted in writing by the author. In the case of an artistic work it may also be asserted by the identification of the author on the original, or copy of the work, or on a frame, mount or anything else to which it is attached.

A treatment of a work is derogatory if it amounts to a distortion or mutilation of the work or is otherwise prejudicial to the honour or reputation of the author.

These rights do not apply to certain works published in newspapers, magazines, encyclopaedias or other collective reference works. Nor do they apply to any work made for the purpose of reporting current events.

Generally, these rights do not apply to employee's works where, as is usually the case, the copyright originally vests in the employer. As regards derogatory treatment, however, the right will apply if the employee has been identified as the author and a 'sufficient disclaimer' is not published.

There are also rights relating to the false attribution of work and to safeguarding the privacy of photographs commissioned for private and domestic purposes. Anyone entitled to these rights may waive them either generally or by reference to specific works, and either conditionally or unconditionally. Any waiver must be in writing.

Legal deposit (Publisher's) copies

Under the *Copyright Act 1911* the publisher of any book published in the United Kingdom must, within one month after publication, deliver at his own expense a copy of the book to the British Library. The copy should be sent to: The Legal Deposit Office; The British Library, Boston Spa, Wetherby, West Yorkshire LS23 7BY. Copies of newspapers should be sent to: The Newspaper Legal Deposit Office, The British Library, Unit 3, 120 Colindale Avenue, London NW9 5LF. Enquiries can be made to the Legal Deposit Office on 071-323 7382.

The copy deposited must be a copy of the whole book, all maps and illustrations belonging thereto, finished and coloured in the same manner as the best copies of the book are published, and bound and on the best paper on which the book is printed.

The publisher must also, if written demand is made within twelve months from the date of publication, deliver, to such address as is named, within one month of such demand, a copy of any book to each of the following libraries - the Bodleian Library, Oxford; the University Library, Cambridge; The National Library of Scotland, Edinburgh; the Library of Trinity College, Dublin; and the National Library of Wales, Aberystwyth.

In the case of an encyclopedia, newspaper, review, magazine or work published in parts, the demand may include all numbers or parts of the work which may be subsequently published.

The copy delivered to these libraries must be on the paper on which the largest number of copies of the book is printed for sale, and must be in like condition as the books prepared for sale.

Any publisher failing to comply with these regulations is liable to a fine not exceeding £50 and the value of the book, the fine to be paid to the authority to whom the book ought to have been delivered.

The expression 'book' includes every part of a book, pamphlet, sheet of letterpress, sheet of music, map, plan, chart, or table separately published, but shall not include any second or subsequent edition of a book unless it contains additions or alterations either in the letterpress or in the maps, prints, or other engravings belonging to it.

By a statute in 1932 certain classes of publications were exempted from the requirement as to delivery of copies to the British Library. These classes included publications wholly or mainly in the nature of trade advertisements, local timetables, calendars and registers of voters.

Stationers' Hall Register

Registration under the *Copyright Act 1911* terminated on 31st December 1923, and no registration is now required in order to obtain copyright in literary, dramatic, musical and artistic works. However, provision is made for registration of designs at the Designs Registry.

Also, there is in existence a Register established by the Stationers' Company at Stationers' Hall, Ludgate Hill, London EC4M 7DD, in which 'Books' and 'Fine Arts' can be registered. This register is not a continuation of the former Register and is not kept pursuant of any Statute. The entries made therein are for the purpose of record and for assistance in the proof of the existence of a work on a given date in the case of infringement.

Books, periodicals, film scenarios and synopses, pamphlets, maps, charts, music, business and trade circulars, catalogues, price lists, commercial documents, sheets of letterpress or other compilations, whether published or unpublished, are included under the term 'Book', and paintings, drawings, designs, labels, photographs and engravings, may be registered under the term 'Fine Arts'. 'Registered at Stationers' Hall' may be added to any registered work.

A copy of every work for which registration is desired must be filed with the Registrar at Stationers' Hall, and certified copies of the entries in the Register are issued on request. The charges for registration at time of going to press were: for any item under the terms 'Book' or 'Fine Arts' £20 + VAT. Certified copies of entries supplied at £10 + VAT each. Searches for titles or proprietors' names contained in the Register carried out for a fee of £7 + VAT.

• Harmful publications, such as 'horror' comics

Under the *Children and Young Persons (Harmful Publications) Act 1955*, it is an offence to print or publish any book, magazine or other like work which is of a kind likely to fall into the hands of children or young persons and consists wholly or mainly of stories in pictures (with or without the addition of written matter), being stories portraying:
a) the commission of crimes: or
b) acts of violence or cruelty; or
c) incidents of a repulsive or horrible nature; in such a way that the work as a whole would tend to corrupt a young person into whose hands it might fall.

A person found guilty of this offence is liable to imprisonment for a term not exceeding four months or to a fine not exceeding £1000 or to both.

• Obscene publications

The *Obscene Publications Acts 1959* and *1964* make it a statutory offence for any person to distribute, circulate, sell, let on hire, or lend obscene matter, or to have an obscene article in his possession or control for publication for gain (whether gain to himself or to another). An article is deemed obscene if its effect is, if taken as a whole, such as to tend to deprave and corrupt persons who are likely, having regard to all relevant circumstances, to read, see or hear the matter contained or embodied in it. Also, an article is deemed to be had for publication if it is had for the reproduction or manufacture therefrom of articles for publication, and this would include negatives, plates and similar items.

It is a defence to show that publication of the article is justified as being for the public good on the grounds that it is in the interest of science, literature, art or learning, or other objects of general concern.

• Libel

A libel has been defined as a statement in printing (or other permanent form) concerning any person which exposes him to hatred or ridicule or which tends to injure him in his profession, trade or calling.

The printer is responsible, jointly with the publisher and author, for all libels published in papers printed by him. He makes a profit out of the printing, and he is still regarded as having a duty to make himself acquainted with what is printed in his printing office.

Ignorance on his part will not excuse him, but the *Law Reform (Married Women and Tortfeasors) Act 1935* allows judges to discriminate as between printer, author and publisher in regard to damages. An indemnity against the publication of a libel is not unlawful unless the person indemnified is knowingly a part to the libel's publication.

There are a number of defences to an action for libel, the two principal ones, apart from 'privilege', being 'justification', and 'fair comment'. For a defence of justification to succeed it must be established that the words complained of are true in substance and in fact.

The defence of justification is now subject to the provisions of the *Rehabilitation of Offenders Act 1974*, under which most people convicted of criminal offences are normally entitled to be treated as if they had never committed such offences after a 'rehabilitation period'.

The defence of fair comment, which only protects statements of opinion, is a plea that the words complained of are a comment, based on actual facts, on a matter of public interest and not malicious.

The *Defamation Act 1952* introduced a new defence in cases of unintentional defamation. The Act provides that a person who publishes a libel may offer to publish a suitable correction and apology and if this offer is not accepted it will be a defence to prove that the words complained of were 'published innocently'.

Words are regarded as published innocently in relation to another person if the publisher did not intend to publish them concerning that other person, or the words were not defamatory on the face of them, and the publisher exercised all reasonable care in relation to their publication.

• Imprint

Under the *Newspapers, Printers and Reading Rooms Repeal Act 1869*, as amended by the *Printer's Imprint Act 1961*, a printer is bound to put his name and address on the first or last leaf of every paper or book he prints, under a penalty not exceeding £200 per copy.

Certain specified publications are exempt, viz - Bank notes; bills of exchange; promissory notes; bonds or securities for the payment of money; bills of lading; policies of insurance; letters of attorney; deeds; agreements; transfers or assignments of stocks or shares; receipts; law proceedings; and papers printed by authority of Parliament or of any public board or public officer.

There is also a general exemption in respect of the printing of the name and address, or business or profession, of any person, and the articles in which he deals or the services which he offers; also in respect of any papers for the sale of estates or goods, by auction or otherwise. This appears to embrace address cards, business cards, price lists, and catalogues of goods or property for sale or of services offered.

The 1961 Act exempts papers and books which do not contain 'words grouped together in a manner calculated to convey a message' or 'drawing, illustration, or other picture'. Where the message is purely a conventional one, such as a greeting or invitation, or the picture represents only a design or registered trademark there is no need for the imprint.

The following are examples of items which do not require an imprint provided they are normal items of the kind mentioned and do not bear a message (other than a conventional one) or a picture (other than a design or a trademark):

Stationery sheets, cards, and forms designed to be completed in manuscript or by other means of record, including: letter headings; form letters; forms for use in connection with accountancy; order forms; certificates and declaration forms; proposal and application forms; entry forms for competitions, and so on; medical records and temperature charts; menu cards and place cards; and skeleton bills of quantities.

Stationery books, including exercise books; scrap books; drawing books; of graph paper; account books; registers, log books; record books; report books and letter copying books.

The following items of packaging materials: cartons; paper bags; wrappers; gummed tape; and self-adhesive tape.

The following other items: address books; compliment slips; correspondence cards; invitation cards; albums for stamps; autographs; photographs and the like; labels; tags; seals; envelopes; card index systems, filing systems and requisites; photographic mounts and folders; paper serviettes, paper doylies, paper table covers, paper table decorations; shelf paper and similar articles of paper.

Any agreement to omit the imprint from a work which, under the law, requires the imprint is illegal and void, and in a case decided in 1822 the omission of the imprint - even by desire of the customer - was held to disentitle the printer to recover for the work done, in addition to rendering him liable to the penalty.

Every person assisting in publishing or dispersing a book or paper not bearing the name and address of the printer is also liable to a penalty of £50 per copy.

The old Act was declared not to apply to Ireland. However, the position in Northern Ireland is now governed by the *Printed Documents (Northern Ireland) Act 1970* which contains similar requirements but applies generally to a wider range of articles and there is similar legislation also in the Irish Republic.

Under the *Representation of the People Act 1983*, the name and address of the printer and publisher must appear on every bill, placard or poster having reference to an election and on every printed document distributed for the purpose of promoting or procuring the election of a candidate.

• File copies

Under the *Newspapers, Printers and Reading Rooms Repeal Act of 1869*, as amended by the *Printer's Imprint Act, 1961*, a copy with the name and abode of the customer on it has to be kept for six months of everything printed for reward which is required to bear the printer's imprint, under a fixed penalty of £500.

• Sedition

The *Incitement to Disaffection Act 1934* is directed (Section 1) against persons who maliciously and advisedly endeavour to seduce any member of Her Majesty's forces from his duty to allegiance to Her Majesty.

Section 2, which may intimately concern printers, states that if any person, with intent to commit or to aid, abet, counsel, or procure the commission of an offence under Section 1 of the Act, has in his possession or under his control any document of such a nature that the dissemination of copies thereof among members of Her Majesty's forces would constitute such an offence, he shall be guilty of an offence under the Act.

Search warrants may be used and anything seized may be retained for a period not exceeding one month or until the conclusion of any proceedings commenced within that time. Anyone found guilty on indictment is liable to imprisonment for up to two years and/or unlimited fine.

• Stolen or lost goods

Under the *Theft Act 1968* the offer in a public advertisement of a reward for the return of stolen or lost goods or any words to the effect that no questions will be asked, or that the person producing the goods will be safe from apprehension or inquiry, or that money paid for the purchase of the goods or advanced by way of loan on them will be repaid, will render the person advertising the reward and any person who prints or publishes the advertisement liable on summary conviction to a fine not exceeding £1000.

• Sunday observance

It is an offence to print any notice or advertisement of any Sunday entertainment (with certain exceptions) to which the public pays or buys tickets for admission.

The exceptions are:
a) any cinematograph entertainment at a place which is duly licensed,
b) any museum, picture gallery, zoological or botanical garden or aquarium,
c) any lecture or debate,
d) any musical entertainment at any place licensed to be opened and used on Sundays.

Musical entertainment is defined as a concert or similar entertainment consisting of the performance of music with or without singing or recitation, but this definition excludes anything in the nature of a variety entertainment.

Sex discrimination

Under the *Sex Discrimination Act 1975* as amended by the *Sex Discrimination Act 1986* and the *Employment Act 1989* it is unlawful to discriminate in employment (and certain other fields) on the grounds of sex or against married people on grounds of marital status.

Discriminatory advertisements

It is unlawful to publish, or cause to be published, an advertisement which indicates, or might be regarded as indicating, an intention to discriminate unlawfully. However, it is a defence for a publisher to show that he was relying (and it was reasonable for him to rely) on a statement by the advertiser that the intended act would not be unlawful.

Race relations

Restrictions on press freedom

The *Public Order Act 1986* provides that it is a criminal offence for a person to publish, distribute, or have in his possession for publication or distribution, any 'written material' (see below) which is threatening, abusive and insulting, if either:

a) he intends to stir up hatred against a racial group (whether it is likely to be stirred up or not); or
b) hatred against a racial group is likely to be stirred up (whether he intended it to be stirred up or not).

In a case where it cannot be shown that the defendant intended to stir up hatred, it is a defence for the accused to prove that he was not aware of the contents of the written matter and neither suspected nor had reason to suspect it of being threatening, abusive or insulting. Also, the offence does not extend to fair and accurate reports of judicial proceedings (if contemporaneous) and Parliamentary proceedings. For the purpose of the Act, 'written material' includes any sign or other visible representation.

Banknote reproduction

The reproduction of banknotes in England, Wales, Scotland and Northern Ireland is governed by the *Forgery and Counterfeiting Act 1981*.

Section 18 of the 1981 Act makes it an offence, unless the relevant authority has previously consented in writing, to reproduce on any substance whatsoever, and whether or not on the correct scale, any British currency note or any part of a British currency note.

'Relevant authority' means the authority empowered by law to issue the note. The main note-issuing authority is the Bank of England, which also claims copyright in its notes.

Guidelines have been issued by the Bank for reproducing its notes in advertisements and illustrations. The Bank restricts reproductions of its notes because it needs to ensure that reproductions cannot be used to defraud the public. It is also concerned about the context in which its notes are reproduced, especially if the Queen's portrait is distorted or shown in offensive surroundings.

Correspondence about the reproduction of notes should be addressed to: The Principal, Issue Office, Bank of England, Threadneedle Street, London EC2R 8AH.

Postage stamp reproduction

No postage stamp or major part of the design of any postage stamp may be reproduced without the written authority of the Post Office. Letters requesting authority should be addressed to Royal Mail Stamps and Philately, Room 201, 76/86 Turnhill Street, London EC1M 5NS (Tel 071 250 2026).

The reproduction of stamps will normally be authorised for philatelic and educational purposes. Other requests will be considered on their merits and there may be a charge where the reproduction is put to commercial use.

Postmark reproduction

No postmark may be reproduced without the written authority of the Post Office. Letters requesting authority should be addressed to Royal Mail Stamps and Philately, Room 201, 76/86 Turnhill Street, London EC1M 5NS (Tel 071 250 2547).

The reproduction of postmarks will normally be authorised for philatelic, educational, decorative and most advertising purposes.

Queen's Awards for exports, technology and environment

It is an offence under the *Trade Descriptions Act 1968* for any person in the course of trade or business to use, without proper authority, any device or emblem signifying the Queen's Award or any close resemblance of such a device or emblem as is likely to deceive.

Details of the Award scheme may be obtained from the Secretary, The Queen's Award Office, Dean Bradley House, 52 Horseferry Road, London SW1 (Tel 071 222 2277).

Food labelling and packaging

The *Food Labelling Regulations 1984* (SI 1305) (as amended) set out the basic labelling requirements for most foods when ready for delivery to the consumer or a catering establishment.

They require all food, subject to certain exceptions, to be marked or labelled with:

a) name of food, that is, the customary name unless a name is prescribed by law, in which case that name must be used;
b) list of ingredients (in descending order by weight);
c) an indication of minimum durability (that is, date stamped);
d) any special storage conditions or conditions of use;
e) the name and address of the manufacturer or packer or EC seller;
f) in certain cases, the place and origin of the food; and
g) in certain cases, instructions for use.

Where special emphasis is placed on the presence or low content of a particular ingredient, an indication of the minimum or maximum percentage respectively of the ingredient must be given.

Generally, the particulars must appear on the packaging, on a label attached to the packaging, or on a label that is clearly visible through the packaging. They must be easy to understand, clearly legible and indelible and marked in a conspicuous place in such a way as to be easily visible. The information must not be hidden, obscured or interrupted by any other written or pictorial matter.

Where, under the *Weights and Measures Acts* a food is required to be marked with an indication of net quantity, that indication is to appear in the same field of vision as the name of the food and the indication of minimum durability (if required).

Small packages (that is, where the largest surface has an area of less than 10 square centimetres) need only show the name of the food and, if required, the minimum durability.

Foods which need not bear a list of ingredients include non-processed or non-treated fresh fruit and vegetables; carbonated water to which only carbon dioxide has been added; vinegar derived solely from fermentation of a single product; certain kinds of cheese, butter, fermented milk and fermented cream; flavourings; single ingredient foods; and drinks with an alcoholic strength of more than 1.2% by volume.

Claims that a food has tonic properties, or that it is equivalent to or superior to the milk of a healthy mother, must not appear in the labelling or advertising of the food (though the word 'tonic' may be used in relation to certain soft drinks).

Other claims may only be used subject to specific conditions; these are claims relating to nutritional requirements; suitability for diabetics; slimming or weight control; medical properties; protein value; vitamin or mineral content (unless the food consists of either vitamins or minerals or both or any of these plus carrying agents, or other substances if vitamins or minerals are sold in tablet, capsule or elixir form); polyunsaturated fatty acids; cholesterol content; energy-giving properties; or claims which depend on another food.

Further, the following words or descriptions must not be used unless specific conditions are fulfilled: 'butter'; 'cream'; 'dietary' or 'dietetic'; 'fresh', 'garden' or 'green' (in relation to peas); 'milk'; 'starch-reduced'; 'vitamin'; 'alcohol-free'; 'de-alcoholised'; 'non-alcoholic'; 'shandy'; or 'shandygaff' (if used alone) or 'ginger beer shandy'; 'sweetened liqueur'; 'tonic wine'; 'vintage Scotch whisky'; 'Irish whiskey'; 'blended Scotch whisky' and 'blended Irish Whiskey'. Also, descriptions incorporating names of food or pictorial representations of food, which imply that a food has a particular flavour, must not be misleading.

• Food packaging

The *Materials and Articles in Contact with Food Regulations 1987* (SI 1523) apply to materials and articles which are in their finished state and are intended to come into contact with food. They have been modified in part by the *Food Safety Act 1990 (Consequential Modifications) (No 2) (GB) Order 1990*.

Under the regulations, food contact materials and articles must be manufactured in accordance with good manufacturing practice, that is, in such a way that, under normal or foreseeable conditions they will not transfer their constituents to food in quantities which could endanger human health or cause a deterioration in the organoleptic characteristics of the food or an unacceptable change in its nature, substance or quality. It is an offence to sell, import or use in a food-related business any such material or article which does not comply with this requirement.

Certain particulars must be shown with the product when sold or imported in its finished state but not yet in contact with food. These particulars are:

a) the words 'for food use' or a specific indication of the particular use for which it is intended or the prescribed symbol unless, in the case of retail sale, it is by nature clearly intended to come into contact with food;

b) any special conditions to be observed when it is being used; and
c) either the name or trade name and address (or registered office) or the registered trade mark of the manufacturer or processor or of a seller established with the EEC.

These particulars must be shown clearly, legibly and indelibly on the product itself, or on its own packaging, or by a label affixed to its own packaging. Alternatively, in the case of a retail sale, the particulars may be shown on a sign within the immediate vicinity of the product and clearly visible to purchasers (though the particulars mentioned in (c) above may appear on such a sign only if it is not reasonably practicable for them to appear on the product itself or its label or packaging). In the case of non-retail sales and importations, the particulars may be shown in the accompanying documents. It is an offence to sell by retail any material or article intended for contact with food unless it is so marked.

The use of the expression 'for food use' or similar wording, is restricted to products which have been manufactured in accordance with the regulations. Advertisements for products which do not comply with the regulations are prohibited, but it will be a defence for a publisher to show that he received an infringing advertisement in the ordinary course of business and did not know and had no reason to suspect that it was unlawful. There is also a general 'due diligence' defence to any prosecution under the regulations, the defendant having to show that he took all reasonable precautions and exercised all due diligence to avoid the commission of an offence. The regulations also control the use of vinyl chloride in food contact materials.

• Weights and Measures

The *Weights and Measures Act 1985* requires, among other things, a wide range of pre-packed goods to be marked with indications of quantity and specify the units of measures which it is permissible to use. Some pre-packed goods are required to be sold in specific quantities.

Indication of quantity must be marked in accordance with *Weights and Measures (Quantity Marking and Abbreviations of Units) Regulations 1987* (SI 1538).

The indication must:

a) be easy to understand, clearly legible and indelible;
b) be easily visible to the intending purchaser under normal conditions of purchase;
c) not to be in any way hidden, obscured or interrupted by any other written or pictorial matter;
d) if not on the actual container or on a label securely attached to it, be so placed that it cannot be removed without opening the container.

Requirements about the minimum size of characters and authorised abbreviations of units of measurement are also set out in the regulations, as are restrictions on the units which may be used in certain cases.

The Act also provides a system of quantity control over packaged goods known as the 'average system'. In conjunction with the Weights and *Measures (Packaged Goods) Regulations 1986* (SI 2049), as amended by SI 1987/1538, it makes provision for the 'e' mark (an EEC mark in prescribed form) to be affixed to packages in the same field of vision as the nominal quantity.

The 'e' mark is a guarantee by the packer or importer that packages have been made up in accordance with the 'average system' and acts as a 'metrological passport' through the whole of the EEC.

Origin marking of exported print

Firms producing printed matter for export should bear in mind that some importing countries require it to be marked with the country of origin. Some countries prohibit the importation of goods bearing any markings which might mislead as to their origin, such as the use of a language which is not that of the country of manufacture. In such cases the goods are normally required to be clearly marked to show their true origin. Some countries have very complex requirements.

In all cases, however, printers are advised to check the position with their customers (through their agents where appropriate) and the up-to-date situation for a particular country should be checked with the DTI's overseas trade divisions.

General

Many printers use a form of marking which is framed to meet the requirements of our own law as to the showing of the imprint and also the requirements of those foreign countries which require an indication of place of origin. A usual form or working in such cases is 'Printed in England by John Smith and Company, 400 Fleet Street, London, England'.

Lotteries and competitions

It is an offence to print tickets (or other matter, such as advertisements or list of prize-winners) relating to an unlawful lottery. A scheme is regarded as a lottery where participants stand to make a gain through the operation of chance and the participants have subscribed to the prize fund. There has been no case where a purely gratuitous distribution of chances has been held to be a lottery.

Schemes which have been held to constitute lotteries include raffles, sweepstakes, prize competitions where the element of skill is negligible or not decisive, snowball and chain letter ventures and certain kinds of trading gimmicks.

The *Lotteries and Amusements Act 1976* provides that, subject to certain conditions, the following kinds of lotteries are not unlawful: lotteries incidental to exempt entertainments; private lotteries; society lotteries; and local lotteries. Conditions which must be fulfilled are set out in the Act supplemented in relation to 'society lotteries' and 'local lotteries' by the *Lotteries Regulations 1977* (SI 1977 No 256), as amended by the *Lotteries (Amendment) Regulations 1981* (SI 1981 No 109) and *1988* (SI 1988 No 2161). Certain monetary limits specified in the Act have been modified by the *Lotteries (Variation of Monetary Limits) Orders 1981* (SI 1981 No 110) and *1989* (SI 1989 No 1218).

An 'exempt entertainment' means a bazaar, sale of work, fete, dinner, dance, sporting or athletic event or other entertainment of a similar character. Where a lottery is incidental to such an entertainment, it is not unlawful provided none of the prizes in the lottery is a money prize, tickets in the lottery are not sold and the result of the lottery not declared except on the premises on which the entertainment takes place and during its progress, and the lottery is not the only substantial inducement to persons to attend the entertainment. The whole proceeds of any exempt entertainment (including the lottery) must be devoted to purposes other than private gain after deducting expenses, including up to £50 for the purchase of prizes.

The lotteries with which printers of lottery tickets are mainly concerned, however, are *private lotteries, society lotteries* and *local lotteries* (which are lotteries promoted by local authorities. Tickets for such lotteries are required to show certain information (see specimen tickets on page 78).

Prize competitions

It is unlawful to conduct in or through any newspaper (defined as including any journal, magazine, or other periodical publication), or in connection with any trade or business or the sale of any article to the public, prize competitions of the 'forecasting' variety and other competitions success in which does not depend to a substantial degree on the exercise of skill. There is a saving for a pool betting operation carried on by a person whose sole trade or business is that of a book-maker, but this saving does not extend to any such operations carried on through a newspaper.

(1)

(number) Fairway Golf Club, Bunker, Lincs. **Christmas Draw 1994** Counterfoils and unsold tickets must be returned to the promoters by Wednesday, 14th December, 1994.	Printed by G. Ball, The Drive, Bunker, Lincs.	Fairway Golf Club *(number)* Bunker, Lincs. **Christmas Draw 1994** Tickets 25p each Promoters: J. Tee, 6 The Rough, Bunker, Lincs; A. Driver, Forelands, Honour Green, Lincs. Ist Prize: £100 2nd Prize: A turkey 3rd Prize: A goose The sale of tickets in this draw is restricted to members of the Fairway Golf Club. No prize won in this draw shall be paid or delivered by the promoters to any person other than the person to whom the winning ticket was sold by them.

(2)

(number) Fairway Golf Club, Bunker, Lincs. **Christmas Draw 1994** Counterfoils and unsold tickets must be returned to the promoters by Wednesday, 14th December, 1994.	Printed by G. Ball, The Drive, Bunker, Lincs.	Fairway Golf Club *(number)* Bunker, Lincs. *Registered with Print Town District Council* **Christmas Draw 1994** Tickets 25p each Promoters: J. Tee, 6 The Rough, Bunker, Lincs; A. Driver, Forelands, Honour Green, Lincs. Ist Prize: £1000 2nd Prize: £500 3rd Prize: £200 The draw will take place in the clubhouse at 8.30 pm on Saturday, 17th December, 1994.

(3)

(number) Print Town District Council **Christmas Draw 1994** Counterfoils and unsold tickets must be returned to the promoters by Wednesday, 14th December, 1994.	Printed by G. Ball, The Drive, Bunker, Lincs.	Print Town District Council **Christmas Draw 1994** Tickets 25p each Promoters: Print Town District Council Ist Prize: £750 2nd Prize: £400 3rd Prize: £200 The draw will take place in the Town Hall, Print Town at 8.00 pm on Saturday, 17th December, 1994.

(1) *Private lottery: specimen ticket*
(2) *Society lottery: specimen ticket*
(3) *Local lottery: specimen ticket*

• Red Cross and similar emblems

Under the *Geneva Conventions Act 1957* it is unlawful for the emblem of a red cross on a white ground, or the designations 'Red Cross' or 'Geneva Cross', to be used for trade, business or any other purposes without the authority of the Army Council. Generally speaking, such authority is confined to the military forces of the Crown and to duly authorised Voluntary Aid Societies. Further it is unlawful to use the red crescent moon and red lion and sun emblems on a white ground, or the designations 'Red Crescent' and 'Red Lion and Sun', without the authority of the Army Council. Any application for approval to use the red cross or the words referred to should be made, prior to their use, to the Ministry of Defence C2 (AD), Old War Office Building, Whitehall SW1.

Under the Act it is also unlawful to use for the purposes of trade or business or for any other purposes, without the authority of the Board of Trade, any design consisting of a white or silver cross on a red ground, or any design which is a colourable imitation of this design, or any design which is a colourable imitation of the Red Cross or the other emblems and designations referred to in the previous paragraph or any words so nearly resembling the words Red Cross or Geneva Cross as to be capable of being understood as referring thereto.

Royal images and emblems

The Lord Chamberlain's Office has issued guidance on how royal images may be used for commercial purposes. 'Royal images' means photographs, portraits, engravings, effigies and busts of the Queen and members of the Royal Family. In addition, there are certain offences relating to the use of royal emblems or false representations of royal approval; and the British Code of Advertising Practice prohibits portrayals or references to living persons in advertisements without their prior consent, though there are certain exceptions.

Lord Chamberlain's Rules

Stationery - Royal images may be used on the following products provided they are free from advertisement:
- Portrait prints.
- Picture postcards (which may include younger members of the Royal Family in family groups).
- Greetings cards that carry only formal greetings.
- Calendars (as a special exception trade calendars many carry the name of the firm and its trading description).

Advertising - Except when advertising a book, newspaper, magazine article or a television documentary about a member of the Royal Family, royal images may not be used for advertising purposes in any medium. A firm's advertisement may not include photographs of the Royal Family visiting their works or exhibition stands or being publicly involved with their products. Publication of film or photographs of such visits may, however, be used for house journals or for specifically in-house purposes.

Copyright - Any question of copyright involved in the reproduction of a Royal image must be settled by the prospective user direct with the copyright holder. Nothing in the rules gives any right to the use of a particular image. In case of doubt about the application of these rules, reference should be made to the Comptroller, Lord Chamberlain's Office, St James's Palace, London SW1.

• Printer's liability for customers' property

A printer who accepts goods in custody from another person is a bailee of the goods. The liability of the bailee depends in the first instance on the conditions on which he accepts custody of the goods, and many printers include a clause in their contracts to the effect that the customer's property and all property supplied to the printer by or on behalf of the customer will be held and carried at the customer's risk. This is covered in the Standard Conditions of Contract issued by the BPIF (see page 85, condition 11a).

Where no such condition applies, a bailee is under a duty to take reasonable care of the goods bailed to him and if the goods are damaged or lost the printer must prove that the damage or loss was not the result of his negligence.

One of the matters to which a court is likely to have regard in deciding whether the printer-bailee has been negligent is whether he is being paid for his services (or is receiving a benefit in some other way), as the duty of reasonable care is put somewhat higher in the case of a bailee for reward than in the case of a gratuitous bailee.

The use of a condition to the effect that the customer's property is held at the customer's risk will protect the printer-bailee in most cases where the property is lost or damaged but will not usually protect him where the loss or damage in the result of a fundamental breach of his part such as the deliberate destruction of the goods, storing the goods in a place different from the place agreed, or unreasonably retaining possession of the goods after their return has been demanded by the customer.

The *Unfair Contract Terms Act 1977* provides, among other things, that a person (which includes a company) cannot by reference to a contract term exclude or restrict his liability for negligence resulting in loss or damage to goods except insofar as the term satisfies a test of 'reasonableness'.

The same test also applies to contract terms which exclude or restrict liability for breach of contract - though in this case, as far as dealing with other businesses is concerned, the control by this test applies only if the clause is included in a firm's written standard terms of business.

In view of this new legislation the question is sometimes raised as to whether the clause in the BPIF's Standard Conditions of Contract is legally enforceable. The test of 'reasonableness' in these cases is whether the clause was a fair and reasonable one to be included having regard to the circumstances which were, or ought reasonably to have been, known to or in the contemplation of the parties when the contract was made.

The Standard Conditions, (see pages 82 to 87), were not issued in their current form until all the objections of the Office of Fair Trading, with which drafts of the document were discussed, were met. It is therefore felt that by objective standards the clause is reasonable and that in relation to any particular case it would be justified on grounds that (a) printers have contracted on this basis for many years and customers are, or should be, well aware of the position; and (b) the clause enables both parties to avoid the unnecessary extra expenses of double insurance. However, in the absence of test cases and without reference to the particular facts of each case, it is not possible to be more specific.

If the printer supplied the paper or other materials necessary for an order, these materials are at the printer's risk until the ownership thereof has passed to the customer. The time when the ownership passes depends on the terms of the contract but a condition used by many printers provides that ownership passes on delivery or on notification that the work has been completed. If in such a case the printer remains in possession of goods after having notified the customer that the work has been completed he is then a bailee of the goods.

• Printer's rights over customers' property

A printer has a particular lien, that is, a right to retain possession, over a customer's goods if a debt is due to the printer for the manufacture or processing of those goods. A printer does not have a general lien, that is, right over all the customer's goods in his possession, unless this has been expressly provided for in his contract with the customer.

Many printers include a term in their conditions of sale which gives them a general lien and also entitles them after due notice to dispose of a customer's goods and apply any proceeds towards any debt from that customer. The Standard Conditions of Contract, (see page 86, condition 13b), contain a clause giving the printer a general lien in cases where the customer is insolvent.

• Ownership of plates and other intermediates

The ownership of plates which a printer makes or has made in the course of printing a job for a customer depends, in the absence of any specific agreement between the parties, on the contents of the relevant documents in each individual case, including the estimate, the order, and any correspondence which has been exchanged. The customer is entitled to the plates only if he has been specifically charged for them (but not if he has been charged only for the making of plates). The customer is not entitled to the plates, even if their cost has been covered in the printers' charge, unless the printer in his invoice, quotation or elsewhere has indicated that his charge includes the supply of the plates. Thus an estimate or invoice which was merely worded 'To printing 1000 copies of 4pp leaflet, £250' would give the customer no right to any plates made and used by the printer in producing the leaflets.

Where the customer is specifically charged for the making of plates but not for the plates themselves, the ownership of the plates remains with the printer but the printer would not be able to reproduce the image on them without the customer's authority.

The same principles apply to negatives or positives, blocks, stereos, electros, moulds, dies, gravure cylinders, tapes, discs and other intermediates.

• Standard Conditions of Contract

In June 1976 the BPIF issued a set of *Standard Conditions* for use by member firms. The BPIF had issued standard conditions for many years prior to 1965, but in that year the existing conditions were withdrawn in view of legislation on restrictive trade practices, though many printers continued to use the same or similar conditions as their own conditions of sale. The new standard conditions were issued after discussions with the Office of Fair Trading and are printed overleaf.

For the standard conditions to have full legal force in any individual case it is essential that they should be drawn to the customer's attention.

It is not sufficient merely to print the conditions on the back of the estimate form: there must also be wording on the face of the form making reference to the printing on the back. This wording must be sufficiently prominent to prevent a customer alleging that he did not see it, and it is suggested that it should be printed as part of the estimate letter (that is, above the signature) in preference to printing it at the foot of the form.

The wording could be on the following lines:

'This estimate is given subject to the Standard Conditions of Contract issued by the British Printing Industries Federation and printed overleaf which conditions shall be deemed to be embodied in any contract based on or arising out of this estimate except as may be otherwise indicated herein or subsequently agreed in writing.'

Where a firm wishes to use its own special conditions in addition to the standard conditions (or some of them), these special conditions may be printed on the face of the estimate, preferably above the signature, or on the reverse of the estimate underneath the standard conditions provided it is made clear that the additional clauses are not part of the standard conditions.

Firms may also wish to consider including a further statement on the face of the estimate form on the following lines:

'Unless previously withdrawn this estimate is open for acceptance within ... days from the date thereof.'

The purpose of this is to save the printer from the possible difficulty that may occur where a customer accepts an estimate at a later date when the printer's production programme has been filled with other orders.

The unconditional acceptance by a customer of a printer's estimate constitutes a contract on the printer's conditions. Some customers, however, specify in their orders certain conditions of their own. A conditional acceptance of a printers' estimate is not binding on the printer until he confirms or accepts it in writing or by conduct. If the printer accepts such a counter-offer the conditions it contains become added to or (where in conflict) substituted for those in the printer's original estimate.

Bromides of the conditions in a size suitable for use on A4 paper (type area, including heading, 270mm x 180mm) are available to Federation members and may be ordered from the BPIF. The bromide text of the conditions is set in Univers 8 on 9pt, arranged in two columns, with paragraph headings in matching bold.

Full explanatory notes on the standard conditions are available from the BPIF to its members. These notes should assist firms in deciding whether the standard conditions are adequate in all respects for their own businesses.

The Standard Conditions of Contract (as shown below), together with the variations issued for continuous stationery manufacturers, carton manufacturers, binders and print finishers and machine readable codes, are registered with the Office of Fair Trading and have been cleared under Section 21(2) of the *Restrictive Trade Practices Act 1976.*

Standard Conditions of Contract as issued by the British Printing Industries Federation

- where appropriate for 'printer' read 'binder'

1) *Price variation* - Estimates are based on the printer's current costs of production and, unless otherwise agreed, are subject to amendment on or at any time after acceptance to meet any rise or fall in such costs.

2) *Tax* - Except in the case of a customer who is not contracting in the course of a business nor holding himself out as doing so, the printer reserves the right to charge the amount of any value added tax payable whether or not included on the estimate or invoice.

3) *Preliminary work* - All work carried out, whether experimentally or otherwise, at the customer's request shall be charged.

4) *Copy* - A charge may be made to cover any additional work involved where copy supplied is not clear and legible.

5) *Proofs* - Proofs of all work may be submitted for customer's approval and the printer shall incur no liability for any errors not corrected by the customer in proofs so submitted. Customer's alterations and additional proofs necessitated thereby shall be charged extra. When style, type or layout is left to the printer's judgement, changes therefrom made by the customer shall be charged extra.

6) *Delivery and payment* - a) Delivery of work shall be accepted when tendered and thereupon or, if earlier, on notification that the work has been completed, the ownership shall pass and payment shall become due.
b) Unless otherwise specified the price quoted is for delivery of the work to the customer's address as set out in the estimate. A charge may be made to cover any extra costs involved for delivery to a different address.
c) Should expedited delivery be agreed, an extra may be charged to cover any overtime or any other additional costs involved.
d) Should work be suspended at the request of or delayed through any default of the customer for a period of 30 days, the printer shall then be entitled to payment for work already carried out, materials specially ordered and other additional costs including storage.

7) *Variations in quantity* - Every endeavour will be made to deliver the correct quantity ordered, but the estimates are conditional upon margins of 5% for work in one colour only and 10% for other work being allowed for overs or shortage (4% and 8% respectively for quantities exceeding 50 000) the same to be charged or deducted.

8) *Claims* - Advice or damage, delay or partial loss of goods in transit or of non-delivery must be given in writing to the printer and the carrier within three clear days of delivery (or, in the case of non-delivery, within 28 days of despatch of the goods) and any claim in respect thereof must be made in writing to the printer and the carrier within seven clear days of delivery (or, in the case of non-delivery, within 42 days of despatch). All other claims must be made in writing to the printer within 28 days of delivery. The printer shall not be liable in respect of any claim unless the aforementioned requirements have been complied with except in any particular case where the customer proves that: a) it was not possible to comply with the requirements, and b) advice (where required) was given and the claim made as soon as reasonably possible.

9) *Liability* - The printer shall not be liable for any loss to the customer arising from delay in transit not caused by the printer.

10) *Standing material* - a) Metal, film, glass and other materials owned by the printer and used by him in the production of type, plates, moulds, stereotypes, electrotypes, film-setting, negatives, positives and the like shall remain his exclusive property. Such items when supplied by the customer shall remain the customer's property.
b) Type may be distributed and lithographic, photogravure or other work effaced immediately after the order is executed unless written arrangements are made to the contrary. In the latter event, rent may be charged.

11) *Customer's property* - a) Except in the case of a customer who is not contracting in the course of a business nor holding himself out as doing so, customer's property and all property supplied to the printer by or on behalf of the customer shall while it is in the possession of the printer or in transit to or from the customer be deemed to be at the customer's risk unless otherwise agreed and the customer should insure accordingly.
b) The printer shall be entitled to make a reasonable charge for the storage of any customer's property left with the printer before receipt of the order or after notification to the customer or completion of the work.

12) *Materials supplied by the customer* * - a) The printer may reject any paper, plates or other materials supplied or specified by the customer which appear to him to be unsuitable. Additional costs incurred if materials are found to be unsuitable during production may be charged except that if the whole or any part of such additional cost could have been avoided but for unreasonable delay by the printer in ascertaining the unsuitability of the materials then that amount shall not be charged to the customer.

b) Where materials are so supplied or specified, the printer will take every care to secure the best results, but responsibility will not be accepted for imperfect work caused by defects in or unsuitability of materials so supplied or specified.
c) Quantities of materials supplied shall be adequate to cover normal spoilage.

13) *Insolvency* - If the customer ceases to pay his debts in the ordinary course of business or cannot pay his debts as they become due or being a company is deemed to be unable to pay its debts or has a winding-up petition issued against it or being a person commits an act of bankruptcy petition issued against him, the printer without prejudice to other remedies shall:
a) have the right not to proceed further with the contract or any other work for the customer and be entitled to charge for work already carried out (whether completed or not) and materials purchased for the customer, such charge to be an immediate debt due to him, and
b) in respect of all unpaid debts due from the customer have general lien on all goods and property in his possession (whether worked on or not) and shall be entitled on the expiration of 14 days' notice to dispose of such goods or property in such manner and at such price as he thinks fit and to apply the proceeds towards such debts.

14) *Illegal matters* - a) The printer shall not be required to print any matter which in his opinion is or may be of an illegal or libellous nature or an infringement of the proprietary or other rights of any third party.
b) The printer shall be indemnified by the customer in respect of any claims, costs and expenses arising out of any libellous matter or any infringement of copyright, patent, design or of any other proprietary or personal rights contained in any material printed for the customer. The indemnity shall extend to any amounts paid on a lawyers advice in settlement of any claim.

15) *Periodical publications* - A contract for the printing of a periodical publication may not be terminated by either party unless 13 weeks notice is given in the case of periodicals produced monthly or more frequently or 26 weeks notice in writing is given in the case of other periodicals. Notice may be given at any time but wherever possible should be given after completion of work on any one issue. Nevertheless the printer may terminate any such contract forthwith should any sum due thereunder remain unpaid.

16) *Force majeure* - The printer shall be under no liability if he shall be unable to carry out any provision of the contract for any reason beyond his control including (without limiting the foregoing) Act of God, legislation, war, fire, flood, drought, failure of power supply, lock-out, strike or other action taken by employees in contemplation or furtherance of a dispute or owing to any inability to procure materials required for the performance of the contract. During the continuance of such a contingency the customer may by written notice to the printer elect to terminate the contract and pay for work done and materials used, but subject thereto shall otherwise accept delivery when available.

17) *Law* - These conditions and all other express terms of contract shall be governed and constructed in accordance with the laws of England.

Notes on Standard Conditions

i) In clause 17, for 'England' substitute 'Scotland' in the case of Scottish printers.

ii) *Continuous stationery* The following clause has been recommended for continuous stationery production in substitution of condition No 7:

'7. *Variations in quantity* - Every endeavour will be made to deliver the correct quantity ordered but quotations are conditional upon the following margins being allowed for overs or shortages (measured in fold depths), the same to be charged or deducted.

For quantities below 10 000, or where special papers or special features are required: 10% margin.

Single-part or one-process work 10 000 to 50 000: 5% margin
- over 50 000: 4% margin.

Multi-part, multi-unit or multi-process work 10 000 to 50 000: 10% margin - over 50 000: 8% margin.'

iii) *Machine readable codes* - The following additional standard condition has been recommended for members who print machine readable codes:

a) In the case of machine readable codes or symbols the printer shall print the same as specified or approved by the customer in accordance with generally accepted standards and procedures.

b) The customer shall be responsible for satisfying himself that the code or symbol will read correctly on the equipment likely to be used by those for whom the code or symbol is intended.

c) The customer shall indemnify the printer against any claim by any party resulting from the code or symbol not reading or not reading correctly for any reason, except to the extent that such claim arises from any failure of the printer to comply with paragraph (a) above which is not attributable to error falling within the tolerances generally accepted in the trade in relation to printing of this sort.'

This additional clause may also be used with the Standard Conditions of Sale for Carton Manufacturers, in which case the words 'printer' and 'customer' should be submitted, wherever they occur, by 'seller' and 'buyer'.

Further Standard Conditions have been issued by the BPIF for *Binders and Print Finishers* and *Carton Manufacturers.*

The use of the male gender has been used throughout the copy in this chapter for reasons of simplicity only and should be considered in the female or male gender as appropriate.

** It is now considered that electronic data supplied to the printer in its various forms comes under the term 'material' and as such is covered by clauses 12a) and b).*

VAT

Value Added Tax (VAT) is chargeable in accordance with the *Value Added Tax Act 1983*, as amended by subsequent *Finance Acts*, and regulations and orders made thereunder. VAT is collected in instalments, with liability to tax arising on each transaction in the chain of production and distribution. Most firms are registered for VAT and are required to account for the tax arising on the goods and services they supply in each accounting period; but they can take credit for the tax arising on goods and services purchased for their business in the same period, and only pay the balance to (or reclaim it from) Customs and Excise.

At the time of going to press the standard rate of tax was 17.5%, but the following items are zero-rated, ie subject to a nil rate of tax:

1) Books, booklets, brochures, pamphlets and leaflets*;
2) Newspapers, journals and periodicals;
3) Children's picture books and painting books;
4) Music (printed, duplicated or manuscript);
5) Maps, charts and topographical plans, and
6) Covers, cases and other articles supplied with items 1 to 5 and not separately accounted for.

** Items which are printed on card or laminated are not regarded as 'leaflets' by the Customs & Excise for VAT purposes.*

Also zero-rated is all advertising by charities for fund raising or educational purposes in newspapers, periodicals, programmes, annuals, leaflets, and similar publications and posters.

Firms are not required to register for VAT if their turnover of taxable supplies is less than £45 000[1] a year; nor are firms supplying only goods or services which are exempted from VAT, though they have to register if they produce their own printed matter to the value of more than £45 000[1] a year.

[1] Total correct at time of going to press - this figure will vary over time due to changes introduced by government.

VAT penalties

The Customs & Excise has power to impose a number of penalties on firms which contravene VAT requirements. For further information a guide *VAT in printing* has been issued by the BPIF. Numerous notices and leaflets are also available from Customs & Excise and these are listed in *VAT on printing*.

6 Production control

As printing is a bespoke industry, making products to a wide range of customers' specific requirements, efficient production control is vital, but difficult to achieve. No one production planning and control system is suitable for all companies - it is therefore necessary to develop a system which meets customer expectations and matches available capacity with production loading.

This chapter covers the following areas: developing a production control system; benefits derived from introducing a production control system; operation of a production control system; monitoring of production data; computerised' production control; long-term and short-term plans.

The type of company structure and main product ranges undertaken makes production control priorities and requirements differ from printer to printer; three main types of printer are identified below to illustrate different approaches.

Long run/'specialist' product printer - eg national magazines, reel-fed labels, cartons, direct mail and books, allowing the opportunity to introduce a system which is sophisticated and with a strong emphasis on reasonably long-term forward planning and operated as a separate management function.

Medium-sized/general printer - eg jobbing-type printer with a mix of work where longer-term forward planning is more difficult. Printing equipment is typically used on a wide variety of products, causing problems in establishing an even and productive work flow. This raises the problem of equipment not being used intensively and being kept standing and interspersed with the opposite problem of bottlenecks. Consideration then has to be given to the balance between 'in-house' or 'bought-in' facilities.

Small-sized/general printer encounters problems similar to the medium-sized printer but normally with the opportunity to be more flexible due to shorter run lengths and quicker turn-round. In this area the printing firm is more concerned with job sequence - ie taking jobs in time or availability order.

Developing a production control system

Some means of controlling production is necessary in all manufacturing industries - printing factories, as with other factories, can benefit considerably from the introduction of a formal, or even an informal system of production control no matter what category of printer they fall into.

Production control can best be described as the management function which *plans, directs* and *controls* the materials supply and processing activities of a company so that production is economic and meets the approved sales/profit target programme, with labour, equipment, materials and capital used to the best advantage. This definition implies that all production control systems are basically the same.

However, this is not the case as each production control system needs to be suitable for the particular application in question; to introduce a complex patented system without first carefully considering total requirements would normally prove a wasteful exercise.

A production control system must be capable of performing the following functions: planning, scheduling, provisioning and control.

Planning is the function which involves short-term activities such as estimating, technical planning, preparation of works instructions, stock control, outwork purchasing and materials control; as well as the provision of information for long-term planning on such matters as sales forecasting, market research, the purchase of capital equipment, finance and labour requirements.

Scheduling involves loading jobs through the factory to meet customers' schedules wherever possible, within the constraints of trying to obtain the optimum and most effective use of available equipment and materials.

Provisioning is the procedure which arranges for the right materials to be in the right place at the right time.

Control ensures that the production plans are being followed, and initiates action when progress is out of step with the plan.

A production control system must be able to maintain and, where possible, increase the profitability of the company; in addition it needs to be capable of meeting the following targets - increase customer satisfaction/service, maintain delivery dates, even out work flow and ensure the highest possible utilisation of facilities. The targets by which the effectiveness of the production control system are judged or measured can vary between one company and another, due to the 'weightings' or priorities given to each area.

It could be that *company (A)* sets its target at achieving as high a level of customer satisfaction as possible - on certain occasions this will only be achieved by interrupting efficient job sequencing or keeping machines idle waiting for late materials such as plates or paper/board. *Company (B)* on the other hand may measure its system's effectiveness by the level of utilisation achieved across labour, equipment and material resources.

Both companies' production control systems aims as outlined can be assessed as diametrically opposed to each other and therefore cannot be easily reconciled. A printing company, by the very nature of the market place it operates within, must be flexible, so achieving the best balance possible of often conflicting aims, as well as maintaining the profit targets set.

An efficient production control system will indicate the productive capacity that is 'loaded' and the total capacity that is available for use. This can be achieved by simply recording the total hours loaded into each load centre or by the more sophisticated method of presenting a visual load picture of the production state of the factory. Data entry can be entirely manual or through a computerised production control module operating within an overall MIS (*Management Information System)*, (see *Chapter 3*).

When planning a production control system the productive capacity of the factory should be stated in the form of a *master diary of parameters*, taken from current production records, which cover all the available resources of the company including shift patterns and working patterns.

One of the key production parameters is *load centres,* which are similar to the cost centres used for costing purposes; however, the load centres may not correspond exactly to the cost centres. It is normal practice for all machines to be load centres, while labour only/handwork operations are grouped into combined centres as appropriate - eg paper or film assembly, bindery benchwork.

Benefits derived from introducing a production control system

An effective production control system should be capable of providing the following functions and benefits:

- indicate the volume of work accepted for each load centre, detailed job-by-job and in total. It will indicate the spare capacity at any date and the type and volume of work required to ensure continuity of maximum production/loading in all departments, enabling delivery dates to be quoted for new and potential orders.

- provide information on the progress of individual orders by showing the extent of work that has been undertaken and the amount carried out on each order in every load centre day-by-day.

- indicate where the load exceeds normal available hours to complete the work in the scheduled time. A decision then has to be taken to arrange extra hours for the job(s) in question by working additional overtime, or where practical, placing the excess capacity with another printing company or suitable trade house.

- highlight/flag up any area where the recorded time or materials used is in excess of the budgeted or estimated figures, enabling investigations to be carried out swiftly. If the extra cost is due to changes which can be attributed to customers - eg author's corrections, then they should be informed accordingly.

- provide a simple basis for assessing at any time the amount and value of work-in-hand.

- be flexible enough to allow sudden re-arrangements due to machine breakdowns, operator's absence through illness or holidays taken at short notice, or accommodate last minute 'rush' orders.

Truly effective production control has jurisdiction over the quantity and variety of stock materials held and the replenishment of stocks before they become exhausted; it should also be responsible for the maintenance of inventory control over completed or partly-completed items. The maintenance of control over stock items is particularly important if the provisioning of the factory is to be carried out effectively.

Operation of a production control system

Production control, in common with other administrative functions, requires documentation. The following are representative of the main documents required by a printing company:

1) **Works order set** consisting of - works instruction ticket; cost sheet; materials release and cutting slip; job control card; plus such other copies as are necessary.

2) **Loading cards**

3) **Stock control cards**

4) **Production programme sheets/work-to lists**

5) **Production return/feedback sheets.**

The documentation used will differ depending on the requirements of the user and whether the system is manually or computer controlled. If an entirely manual system is used all the documentation is prepared by hand and/or through the use of typewriters and word processors.

In many companies the works order set is made up from a carbonless set (see page 25); where companies use remote control data collection/ key pads, production return sheets are dispensed with; a fully-computerised system working as part of an MIS would rely on generating, accessing and manipulating data supported by whatever hard copy documentation was considered appropriate.

In considering the preparation of the works order set, it is anticipated that the preliminary operations of entering the order in the system, informing customers of receipt of the orders, and issuing works order numbers have been attended to.

The next task is to prepare the information for the works order set. If the order is a reprint, the previous works order can be obtained and checked for any possible changes; if correct, it can be retyped on a new works order set.

When the order is new, not a repeat, the first task will be to translate the customer's instructions into technical language. The technical planning can then be carried out and materials requisitioned from stock or ordered as necessary, as well as making adequate provision for any outwork such as artwork, colour separations or plates that may be required.

If the customer has received an estimate, the estimating details will be used as a basis for the preparation of the works instruction set. It is advisable, however, to compare carefully the new order with the estimator's details, and if these differ to any extent, then the estimator should be informed so that customers can be told immediately if the alterations involve any extra charges. When a computerised MIS is operated, an 'electronic' work bag (works instruction ticket) plus other supporting documentation is created by the system as required.

When loading and scheduling the job takes place, the job must be broken down into its component parts and times established for all operations. These can be obtained from the previous order, if the job is a reprint, from the estimate, or, if the job is completely new and not the subject of an estimate, the times can be built up from production data established for this purpose. The hours for the operations may then be entered on the job control card (or its equivalent in the system) as production hours against each load centre.

Job control card

From the production control point of view the job control card is the focal point, carrying all the job details: the front of the card contains the details required to plan production; on the reverse similar details as those on the works instruction ticket can be reproduced.

The customer's name, brief details and job number are normally recorded at the top of the card. The hours for each operation or combination of operations are recorded as production hours. The starting and completion dates show when the operations are planned for, as indicated by the loading board. The row of figures at the top of the cards serves as a diary allowing dates to be checked day-by-day and at regular intervals for 'flagged dates'.

The progress of an individual job through the factory is shown by striking through the relevant completion date, so indicating the job has passed that particular section or department. Alternatively a colour coded system is used - eg *green* - start date of job; *blue* - out on proof/awaiting approval from client; *red* - delivery to customer. *Figure 6.1* illustrates a job control card which has been prepared for a job due to start on the 5th of the month, out on proof from 13th to 15th, and delivery scheduled for the 26th of the month.

Production control is the central point where the information on all aspects of the job is received and distributed. The job control card should be designed so that all this information can be recorded in a logical manner and is readily available to all members of management.

As an alternative to the job control card some companies use 'T' cards, raising only one 'T' card per job or, alternatively, a separate 'T' card for each operation or department.

Visual loading boards

The use of visual loading boards enable the current and future status of jobs to be easily viewed, indicating both the time allowed for the job and sequence in which the jobs will be run. The loading board is divided vertically to represent working periods (that is, single, double or treble shift workings, as appropriate) with segments scaled to represent the working pattern chosen as days and/or hours.

The average output for each load centre must be calculated from the production parameters and records maintained within the company. The load cards can either be pre-printed and carry a graduated scale so that the work content can be marked off, or plain and cut to scale as required.

1	2	3	4	5	6	7	8	9	10	11	12	13	14	15	16	17	18	19	20	21	22	23	24	25	26	27	28	29	30	31

Customer Job no.

Brief details of job

Time/Dates	**Materials Ink/Paper**	**Origination**	**Proofs out**	**Proofs O.K.**	**Plates**	**Press**	**Finishing**	**Delivery**
Production unit								
Production hours								
Start date								
Completion date								

		Cost	**Estimated**
Materials			
Paper			
Ink			
Artwork			
Plates			
Outwork			

Figure 6.1: Example of job control card

The established load centres - such as printing machines - are positioned vertically down the side of the loading board and channelling is fixed horizontally across the board to hold the appropriate load cards. In this way a picture of the work load of the factory can be built up. A movable plastic cursor may be attached to the board so assisting in locating the information required; it is important, however, to note that the loading board should be looked upon as an aid to the system and not as the system itself.

'T' cards

These are T-shaped cards which are placed in racks or slotted panels and are used mainly for *job sequencing.* Essential details such as job number, customer and completion date are normally entered on the visible top bar of the 'T': the remaining part of the card, hidden when placed into the slotted panel, contains additional relevant information related to the job - eg operational times, record of proofs, outwork etc.

'T' cards are often produced as coloured cards, therefore introducing the element of colour coding which significantly improves the visual aspect of the information conveyed by the cards.

A common use of 'T' cards in printing companies is for the production control department to have a *master 'T' card board* such as illustrated by *Figure 6.3* using different coloured cards for the separate departments arranged vertically in job sequence, starting with origination departments and finishing with despatch.

As the job progresses from one department to another, this information is fed back to the production control office where the 'T' cards are removed after the work has been been completed in any particular area. This portrays a visual representation of jobs at any time, showing the sequence in which they stand in each department, as well as indicating outstanding work.

It is often quite common for medium- to large-sized printing companies to operate a *departmental 'T' card system,* separate though complementary to the master production board. The origination department could, for example, cover areas such as artwork, typesetting, paper make-up, film make-up, scanner, proofing, system make-up and platemaking; while the printing and print finishing departments would cover the different machines.

If a secondary 'T' card system is operated within production departments it is essential that there is an efficient *feedback system,* (see pages 99 and 100), such as production return/feedback sheets to keep the production control department always up-to-date.

Printing companies frequently use both visual loading and 'T' card boards together within the one production planning system; in these instances the 'T' card system is used mainly for job sequencing and the visual loading board for machine loading only or to indicate long-term available/booked capacity.

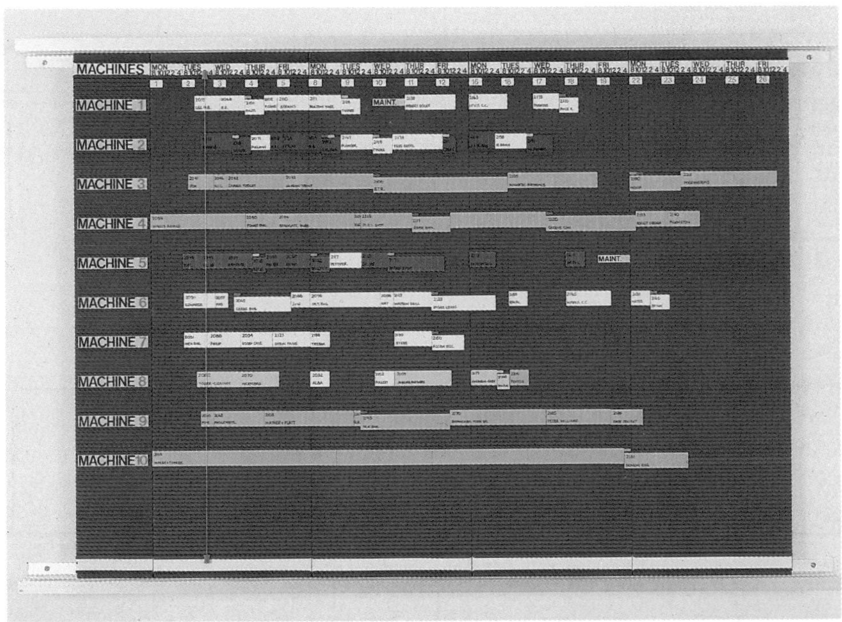

Figure 6.2: Machine loading using visual loading board

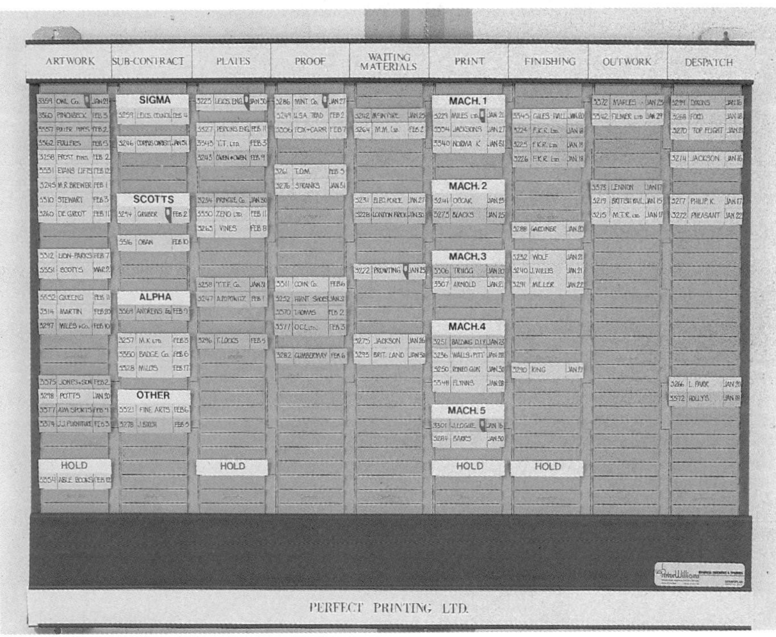

Figure 6.3: Job progressing/sequencing rack system using 'T' cards

Scheduling and loading

Scheduling is the placing of jobs in sequence but it is of little value unless there are adequate arrangements for the productive capacity to meet the schedules agreed upon.

When the various elements of a job are ready for the factory, cards are prepared for each load centre on the loading board required to complete the job in question, with the amount of work marked off to scale.

The load cards are then positioned on the board to meet the job delivery requirements. This implies that the job can be loaded without interruption to other work already planned. This is not always the case and where an overload occurs the question of priorities must be determined. When the load cards for the job have been finally positioned, the starting and completion dates are recorded on the job control card together with the date for delivery.

The success of a production control system is often judged by its ability to arrange the work load to meet the decided and quoted delivery dates. However, many jobs received by a company have no delivery date and these tend to be overlooked in the desire to concentrate on dated work. Ultimately, the undated jobs become urgent purely because they have been in the factory so long.

A well-run production control system should be flexible enough to accommodate undated work, which should be planned and introduced into the system in exactly the same way as the work with an agreed delivery date.

Communication between production control and production departments

The planning carried out in the production control office is of little use unless it is communicated to the people in the sections concerned with the actual production. Similarly, details of the progress made by the factory must be fed back to production control so that effective control can be exercised.

Works instruction ticket

This document will contain the technical planning for the successful execution of the order. The information should be in concise, unambiguous terms which will leave no room for misinterpretation or error, (see page 25).

Materials release and cutting slip

This document, being part of the works order set, will contain brief details of the job, but its main functions are to provide information on the materials requirements, such as paper and board, and to give instructions for cutting the material should this be necessary. It should remain in production control until a pre-determined time before the material is required. This time interval will vary from company to company but it should be no longer than is necessary to issue and prepare the material so that it is ready when the job goes to the machine. If provisioning is carried out correctly, time lost through waiting for material of any sort will be limited, and paper or board will not be left lying around for long periods.

Stock control cards

A separate stock control card should be prepared for every line of stock held, indicating minimum and maximum stock levels, quantity to order and re-ordering levels (see page 34).

Production programme sheets/work-to lists

A production programme is a list of orders in sequence of priority covering the anticipated time it will take for the work load in the programme to be completed. A programme will be issued to individual departments or sections and copies of the programme for the whole factory will be presented to senior managers as required. A programme should be for as long a period as is practical, so it may require amending as the period progresses. A popular procedure is to issue a production programme for the week ahead on a Monday morning, followed by daily or update sheets as the week progresses.

Production return/feedback sheets

In common with most systems, production control would cease to function effectively without a suitable *feedback system.* A simple but effective system is for each production department to present to production control each morning a report of the previous day's production, giving details of completed and partly completed jobs. Production control will note the details on the returns of production. Where work is progressing according to plan no action will be required, but where production is behind plan remedial action will have to be taken.

Production control has a major role in raising the progress of work through a factory from a 'fire-fighting' exercise to one of organised planning. Discipline must prevail at all levels within the company, but this discipline must not be allowed to confine, frustrate or restrict progress but encourage flexibility.

The production controller, being responsible for the production plan, must set the priorities, but when there is conflict over priorities more senior members of the staff should be involved. Properly organised, production control has an important part to play in the management of any modern manufacturing company by planning in the most economical manner and by constantly monitoring the result/action to be taken where necessary. In this way the aim of increased productivity should be achieved.

Monitoring of production data

If a production control system is to be efficient and responsive to complex and changeable production priorities, as typified by most of the printing industry, then it is important that a fast monitoring and control facility exists.

The production return/feedback sheets discussed on page 99 have been found by some printing companies to be difficult to maintain under production pressures over a period of time and so they have sought simpler and quicker response alternatives. For a small-sized, compact company the *'walk-about'* procedure provides an instant overview of how production is progressing against plans. 'Walk-abouts' or visits to production departments can be on a timed interval basis or simply at any time in response to a particular need for information or confirmation of progress.

Shopfloor data collection

In a bid to establish real-time - ie instant access to production data - shopfloor data collection and machine monitoring have been introduced; such systems allow the discontinuation of daily dockets/ time sheets as the system performs the role of time recording.

Shopfloor data collection is basically a system through which production is recorded through *key pads* or *terminals* on the shopfloor with the operators keying in job details as required. The pads or terminals link the production operators to a host unit housed in the production control department where all input data is immediately accessible to the production controller, account executive or other members of management.

Direct machine monitoring

A further facility which has become available to printers in offering real-time benefits is in the area of *direct machine monitoring.*

The system can be simply a counting mechanism which records, for example, the number of sheets printed or folded, and feeds this information back to a host computer which is accessed to provide feedback data.

Alternatively, the system can be part of a sophisticated diagnostic, productivity and monitoring data processing facility such as the MAN Roland 'Pecom', KBA 'Logotronic', Heidelberg 'CP-Data' and Komori 'Plan' systems. Another development in this rapidly expanding area is in companies producing MIS units linking up with machinery manufacturers' monitoring equipment - an example of this collaboration is that between the Optimus MIS and MAN Roland Pecom system.

Computerised production control

Computerised production control in an integrated form is available in two main options, although computers are, of course, constantly used in a variety of ways for inputting and recording data which can be used for production control purposes.

The two main 'dedicated' systems are as follows:

a) production control module linked into a comprehensive MIS; and
b) stand-alone electronic planning and loading board.

MIS-linked production control

In this option the printing company operates an MIS (see *Chapter 3*), where production control works within the overall integrated configuration supported by facilities generated on the system such as estimates, order processing documentation and work-to lists.

Stand-alone electronic planning and loading board

These systems have been developed to take the place of the manually-operated visual loading board driven in most cases from a high-resolution graphics PC.

It performs the functions of manual planning much faster and efficiently as well as providing the following benefits:

- provides clear and informative screen covering of a wide range of production-related data

- calculates an almost limitless range of 'what if' production scenarios
- matches up jobs with available resources
- automates the scheduling of jobs either forward from the date booked in, or backwards from the delivery date
- prepares a wide range of work-to lists covering widely-differing criteria based on, for example, delivery date, most effective utilisation and quickest throughput.

Although the unit has been described as stand-alone it can, in fact, be linked to a host/compatible MIS, so automating the production planning process even further.

Examples of such units are the Optimus '*Optiplan*', '*Optiplan II*' and Shuttleworth '*Printer's Production Scheduling*'.

Long-term and short-term work plans

During the initial stages of working out a schedule for a job with a potential customer, the account executive or production controller works to a *long-term plan*, where target/projected dates and schedules are established. As the time span narrows, reflected in the receipt of the order, or as flagged dates occur, a *short-term plan* is drawn up in much greater and tighter detail.

Short-term plans should be more accurate and specific as the longer the time span the more difficult the prediction.

Print production planning is not just about working out how long an operation will take or where it will most effectively fit into the overall plan, but calculating the time required from receipt of order to completion of a job and delivery to the customer. This involves building in *buffer times* before each of the stages of pre-press, printing and print finishing.

A buffer allowance is the time it takes from the completion of one operation or event to the start of the next one in the production chain. Buffer times will vary depending on the type and throughput of work undertaken by each company, and should include allowances for outwork and customers approving proofs etc.

7 Creating printed images

This is the area where there has been most change in printing organisations - through the adoption of new technology - from previous traditional methods. It is generally recognised that the catalyst to this situation can be traced back to the introduction of word-processors, PCs and, of course, DTP *(desktop publishing)*. 1980 saw the birth of the first personal computers, with the period between 1983 and 1985 seeing the start of DTP, in the launching of the 'Mac' *(Apple Macintosh)* computer followed by the LaserWriter printer which linked up with the Mac.

As software programs developed for the Mac and other PCs, page make-up programs such as *PageMaker* and *QuarkXPress* became available to lift DTP out of its originally intended market of producing internal office reports to mainstream printing and other communication media. The wide success of the Mac is due to its extremely user-friendly interface, WYSIWYG operation and graphics environment, coupled with the ubiquitous PostScript page description language.

To support the growing use of Macs in such a large number of applications, including linking-in with multi-tasking workstations, there is a wide range of models available. In late 1993, the AV *(audio visual)* range was launched as the Centris 660AV and Quadra 840AV.

PCs, under the general heading of 'IBM compatibles', have expanded into more and more powerful units such as the 486DXs and also into a wide range of DTP programs and multi-media facilities. Although Macs are by far the most used computer in printing organisations for DTP work, outside the printing industry, IBM compatibles predominate for DTP and other general business use. Since the launching of Windows *(Microsoft Windows)*, IBM compatibles now have a much more user-friendly environment compared to the previous DOS *(Disk Operating System)* which tended to be cumbersome and more difficult to master compared to the Mac system.

Macs and IBM/PC compatibles are therefore moving closer together in terms of their operating systems and also in the ease of movement of files from one system to the other, through the use of exchange software and compatible software programs. It is likely that Macs and PCs will move closer to an *open platform* basis where compatibility will only become a matter of degree - the joint work between Apple and IBM on the *Power PC* would appear to make this inevitable. This chapter will concentrate mainly on the Mac as the main pre-press input tool, although most, if not all of the areas covered, could be carried out on IBM/PC compatibles.

The specific areas covered include: changing roles of the designer, printer and customer in preparing printed material; pre-press sequence of operations - design brief, sketch/visual, text writing and editing, photography, graphic design, production layout, typesetting, handling disks supplied by clients, proof reading and checking, conventional and electronic reproduction; proofing - presentation of proofs and classification of proofs; platemaking and direct-to-plate systems.

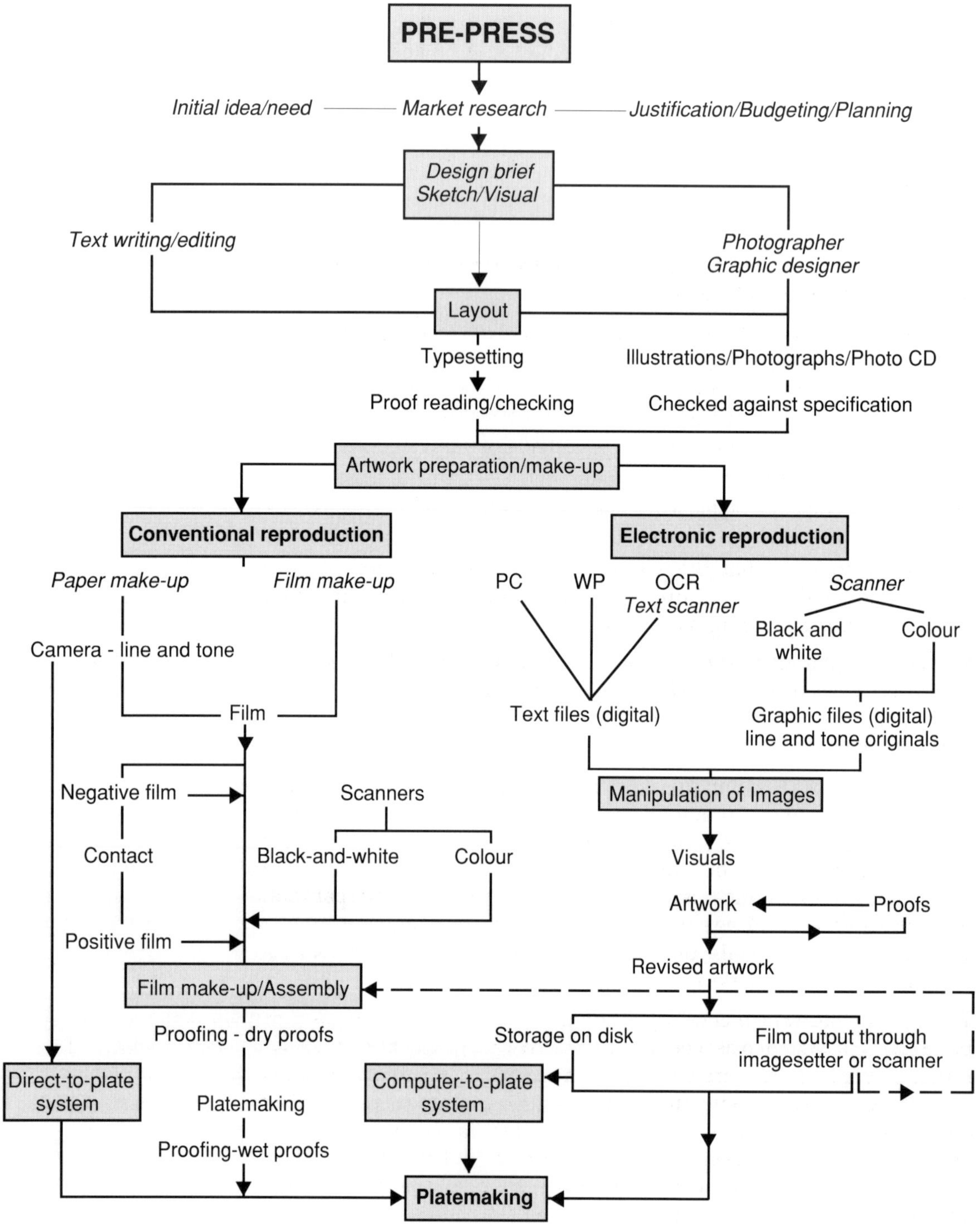

PRE-PRESS

Initial idea/need — *Market research* — *Justification/Budgeting/Planning*

Design brief
Sketch/Visual

Text writing/editing

Photographer
Graphic designer

Layout

Typesetting

Illustrations/Photographs/Photo CD

Proof reading/checking

Checked against specification

Artwork preparation/make-up

Conventional reproduction

Paper make-up

Film make-up

Camera - line and tone

Film

Negative film

Contact

Positive film

Scanners

Black-and-white

Colour

Film make-up/Assembly

Proofing - dry proofs

Platemaking

Proofing-wet proofs

Direct-to-plate system

Electronic reproduction

PC

WP

OCR
Text scanner

Scanner

Black and white

Colour

Text files (digital)

Graphic files (digital)
line and tone originals

Manipulation of Images

Visuals

Artwork

Proofs

Revised artwork

Storage on disk

Film output through imagesetter or scanner

Computer-to-plate system

Platemaking

Figure 7.1: Diagram of pre-press operations outlining conventional and electronic reproduction

Figure 7.1 has been prepared in the form of a flow diagram illustrating the pre-press processes in sequence through conventional and electronic reproduction.

The application of new technology in creating printed images has changed the pre-press aspects of printing practically beyond recognition compared to the previous conventional methods.

Changing roles of the designer, customer and printer in preparing printed material

Traditionally, especially with printed products which require a certain level of design input, it has been normal practice for a graphic designer or design agency to produce the initial sketches, visuals, dummies, layout and possibly make-up finished flat artwork from paper paste-up and airbrush techniques which were then passed onto the printing company for printing.

This long-established, labour-intensive, multi-operational approach is now being seriously challenged by DTP-driven facilities. It is interesting to note that the whole area of pre-press can now be manipulated and prepared on DTP digital data, even up to the final stages of making printing plates.

New technology is relatively accessible to anyone, so that the design agency, printer or the customer can produce directly or through a second party - ie a bureau, the amount of work required, therefore controlling and managing the whole pre-press area.

It is, however, more common for each of the 'players' involved in the *design-and-print cycle* to prepare only part of the pre-press requirements - eg *customers* producing 'copy' on word processors, IBM compatibles or Macs, mainly in the form of disks, which are then passed onto the *graphic designer* to produce visuals for approval, followed by preparation of the finished artwork again in disk/digital form. Finally, the *printer* generates the required film on scanners, imagesetters and workstations for subsequent platemaking, or if applicable goes direct-to-plate.

There are certainly an increasing number of customers handling multi-page publications such as weekly and monthly magazines, journal and report publishers who are controlling the whole pre-press operation up to the stage of supplying fully-formatted PostScript disks or similar data through digital links to a bureau, repro house or printer for outputting on their imagesetter or scanner. Some print buying customers have even installed their own imagesetting and colour scanner/make-up facilities in-house so that final composite film is supplied to the printer.

Pre-press sequence of operations

As printing is largely a bespoke industry the stages of *initial idea/need; market research; justification to proceed, budgeting and planning* up to the *design brief stage* will largely rest with the customer. Once the decision has been reached to produce a printed product, various options are then available.

Design brief - sketch/visual

This stage will normally commence with the customer or customer's representative briefing the person responsible for preparing the initial sketches and visuals. The type of product, its use, house style, corporate image, illustrations, photographs, materials to be used, etc will be discussed at this stage as well as the proposed budget and schedule.

A great many problems and misunderstandings can be eliminated at the outset if preliminary discussions take place between the person responsible for co-ordinating the work for the customer and the printing company's staff. This is to ensure the artwork, film or disk is prepared in the most effective manner and to the correct specification.

The account executive/customer contact within the printing organisation should be included at the earliest opportunity in the job planning meetings, where it should be established what is to be supplied to the printer and in what form, eg camera-ready copy or formatted disks; number and type of illustrations to be used etc. Other areas to be agreed will include the production schedule together with proofing requirements in terms of the types and quantities of proofs with target and approval dates.

Chapter 9 - Print quality control, specifications and standards covers the aspects in print reproduction which need to be considered in ensuring the printed product is to the quality and specification as laid down by the customer and printing company.

The drive towards electronic/digital generation of printed images was reflected by a survey carried out in late 1992, where it was stated that over 70% of the participating design houses and advertising agencies used computers, 85% used them for page layout, 67% for drawing and painting. About half stated they had no intention of bringing colour separation in-house, relying on printers or repro houses to prepare and drop in the colour images using the disks they would supply.

The most popular page make-up software programs used on the Macs for preparing print-related work at the present time are *PageMaker* and *QuarkXPress.*

PageMaker has been easily accepted and integrated into printing companies and typesetting establishments as it follows the pattern of a 'paste-board' frame surrounding the screen page and the traditional method of paste-up by bringing type (and imported graphics) onto the page and placing them in position.

QuarkXPress takes a different route in that pages are first created by placing pre-determined frames in position, into which text and graphics are introduced and then adjusted. Initially, the program was restricted in that frames were created within frames - eg a graphics frame within a text frame and movement could only be within the parent frame, but this procedure has now been simplified.

Both programs have had considerable updates since their original launch - the latest being Aldus PageMaker 5.0 and QuarkXPress 3.3 - each importing text and graphics from other software programs where applicable. QuarkXPress with its strong reference to traditional typographical standards, also its considerable updates and Xtensions, is generally favoured by the professional user; PageMaker, available in Mac and PC versions has more mass market appeal.

Text writing and editing

These are processes which would be carried out either by a member of the customer's staff or handled by a professional copywriter. The text would normally be produced on a word processor or PC and made available to the printer in disk form or clean hard copy suitable for optical character reading.

Photography

Photography undertaken by lay persons will vary considerably in quality depending on the skills of the individual, the equipment used, plus the lighting and subject matter conditions at the time. Professional photographers are often commissioned to take a series of shots to a specific brief of subject matter, composition and balance etc, but the results may still require retouching to ensure the required printed result is achieved. The conventional photographic products used for printing reproduction fall into the main categories of monochrome or colour photographs, colour negatives or positive transparencies.

The application of new technology has impacted on photography to the extent that higher quality digital imaging is now available through the use of *digital cameras* and *Kodak Photo CD.*

Digital cameras produce high-resolution digital imaging which often can be input direct to a computer by plugging into the SCSI port on the back of the camera and downloading images onto the Mac or other SCSI-compatible computer. Other versions of digital cameras capture the images on re-usable and interchangeable video floppy disks.

Kodak Photo CD is another form of digital imaging where up to 100 images from 35mm colour negative or colour reversal film can be stored on a CD master. The Photo CD utilises a proprietary colour encoding system called *Photo YCC*, access software enabling the disk images to be read into the Mac or PC. The Photo CDs are prepared by a specialist photoprocessor. Digital photographic images can be stored in a wide range of media such as Photo CD, magnetic tape, floppy disk and hard disks, making it much more suited to modern electronic printing reproduction.

The quality of digital photography is generally much higher than conventional photographic techniques as traditional photographic emulsions tend to fade and often suffer from colour casts.

Illustrations and photographs when received by the printer should be checked carefully to ensure they are clearly identified and accounted for, also that they are to specification and not damaged in any way. Any damage or problems should be reported to the customer without delay.

Graphic design

Graphic design for print has changed dramatically in the last decade through the application and adoption of CAD *(Computer Aided Design)* in the form of software programs run on Macs and PCs, as well as dedicated high-end design systems which are mainly aimed at specialist areas of printing such as packaging, labels, direct mail and business forms. These systems are now capable of generating complete camera-ready combinations of text and graphics in colour, colour separated, retouched and redrawn to suit particular requirements. The designs are stored in digital form and are therefore easily accessible to change, amend and alter, once the basic design parameters are prepared.

Software design programs are available to suit particular requirements - eg PageMaker and Ventura, for basic type make-up; and Macdraw and Illustrator, for designing, preparing artwork and supporting tools for carton printing and manufacture, continuous stationery etc. It must be acknowledged, however, that conventional manual design techniques with the use of drawing board, air brushing, photography, letraset, and paste-up are still widely used by some graphic designers and possibly a combination of the new CAD and the old traditional creative design techniques produce the best overall results.

A great many problems will be eliminated in the design and production of printed matter by preliminary discussions taking place between the printer and designer to ensure the artwork and/or film is prepared to the correct specification. Good practical design must be allied to a cost-effective, realistic budget.

Production layout

Unless the graphic designer prepares the combined design and image elements on a DTP/CAD system up to the stage of outputting to an imagesetter or other means of output, a *production layout* of each job must be prepared to ensure that the desired printed result is expressed in the elements or components necessary to create the printed images required. The designer will, whatever system is adopted in preparing the work, still need to produce sketches and visuals at the required stages to ensure the customer's approval to proceed and to discuss reproduction aspects with the printer.

Figure 7.2 illustrates the production layout drawn up for the front cover of this publication. Some production layouts are intended to indicate simply the general arrangement of type and illustrations. In these instances just a rough sketch is all that is required, not too elaborate, but giving enough detail to be intelligible to the pre-press department. It should indicate the exact finished size of the job when trimmed, the type area for the text matter and related margins, and the position of display lines, chapter headings, illustrations and captions if applicable.

The typefaces selected should, if possible, be those held by the particular printing company engaged to produce the work, and an indication should be given as to whether the text is to be set solid or line spacing added, also to the treatment of main headings, sub headings etc. As far as the general style of setting for the text matter is concerned, this should follow the printer's house style unless there have been instructions to the contrary.

Production layouts are simply the equivalent of what in engineering would be drawings or blueprints. Coloured pencils, felt tip pens and markers can be used to good effect, indicating the use and split for colour - professionally prepared they can result in high quality production visuals allowing the customer an ideal, inexpensive preview of the finished job. It will be of value to submit such a layout/colour visual to the customer in advance, together with a made-up blank 'dummy' of the job, or a sample of the material to be used, in order to check that the treatment proposed conforms to the customer's requirements and expectations.

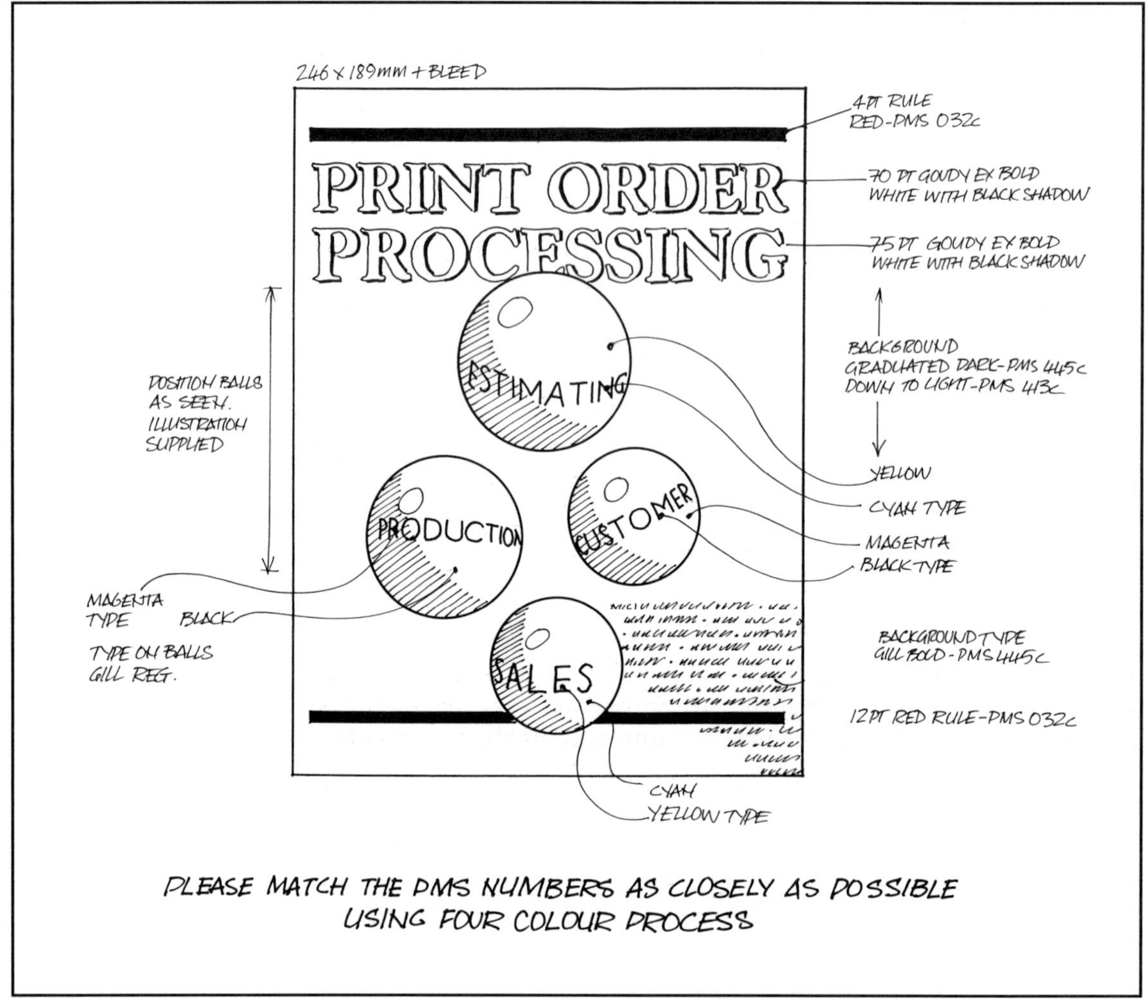

Figure 7.2: Production layout prepared for the front cover of this publication

Typesetting

Traditionally, printing companies have set, or at least organised the setting, of a large proportion of the typesetting element of their printed work. Customers would supply a marked-up manuscript, with the initial keystroking being normally produced as hard copy on a typewriter. This would then be re-keyed by the printer's typesetter with varying degrees of errors creeping in which would need further reading and correcting in a re-keying cycle until the copy was perfectly clean and approved for reproduction.

Nowadays word processors capture the keystrokes in a transferable form - for example as formatted disks - as well as producing a hard typewritten copy for checking, considerably curtailing, if not eliminating entirely in some cases, the need for a printer's typesetter to key again after the customer has created the approved manuscript.

Technology has advanced so rapidly with the advent of DTP and the use of WYSIWYG *(What You See Is What You Get)* terminals that the customer can prepare or interface with the printing company's equipment to produce camera-ready copy. If the customer does not have the original keystrokes captured on a floppy disk, or some other means of digital storage, but only possesses good, clean typewritten/printed hard copy then the use of OCR *(Optical Character Recognition)* facilities will interpret the keystrokes and data scanned from previously typed or printed copy.

OmniPage is a popular OCR software program converting scanned text to MacWrite (text generation application program) or ASCII file format which can be imported into most page make-up software programs and so speed up the process of typesetting.

Handling disks supplied by clients

When representatives of a printing company are discussing with clients what should be supplied to them in the form of disks, the client should be encouraged to adopt the following procedures, as highlighted in italics:

wherever possible, the disks are supplied to the 'ideal' specification for the printing company
- eg 3.5" floppy disks produced on the Mac in PageMaker format. If the disks supplied are incompatible with the printer's system, then the printing company will use conversion facilities which will re-configure the data into a format they can access.

a laser proof of exactly what is on the disk is always supplied
- without this the printer's Mac operator cannot check what is supposed to be on the disk; also, if there are any inconsistencies, 'problem areas' and errors on the hard copy then the printer has the opportunity at an early stage to point this out to the client and establish the best way to resolve the problems.

a complete list of founts used on each job is given
- this is especially important when less well-known typefaces are used, as the operator will need to download these onto the system as and when required.

all disks supplied are marked and labelled clearly as to their contents, with the files to be used identified clearly
- ideally only one file should be on each disk, along with any supporting EPS *(Encapsulated PostScript)* and TIFF *(Tagged Image File Format)* files.

the software program version used to produce the disk is clearly identified - eg PageMaker 5.0 or Quark 3.3. Problems can arise in trying to work between two different versions of the same software program: if, for example, a PageMaker file is supplied to a printing company then assuming that they had the same version or a later one which allowed compatibility with earlier versions, alterations and amendments could be made by the printing company to that file. When the same document is sent as a PostScript file to the printing company they will not be able to access and alter it in any way, although it will contain the EPS and TIFF files required to print the required file.

It is always advisable for a new client to discuss the *do's* and *don'ts* of supplying disks or other forms of digital media in the first instance with the pre-press manager or senior operator of the printing company. Once the initial teething problems have been overcome, account executives should be able to ensure the work is received and supplied in the most efficient and cost-effective manner.

Proof reading and checking

Most good quality word processing/page make-up applications have spell-check facilities; in addition there are now software programs which can check the grammar of text blocks. Copy preparation, by a printing company in its traditional sense, is therefore receding somewhat, but in some ways new technology has heightened the need to maintain high standards of typography and consistency.

Easy and simple access to typesetting preparation through desktop publishing has brought the reward of much cheaper setting, but with it the lack of skill and appreciation which is so necessary to produce good typographical standards. The printing company, although setting less and less copy, is still to a great extent the guardian of those standards.

In most printing companies the role of the reader has changed considerably in recent years, due to the changes to typesetting and image preparation, as outlined in this chapter. The reader/quality checker is now often responsible for checking the accuracy of paper/film make-up, camera-ready copy, plates, ruled-up copies for position, and the final check before *pass-for-press.*

The BPIF publication *Introduction to Printing Technology* gives greater coverage of copy preparation and proof correction, including proof correction marks.

Artwork preparation and make-up can be prepared 'conventionally' or 'electronically' or, in fact, a combination of the two; the remainder of this chapter outlines the approach to both methods.

Conventional reproduction

Up to the stage of platemaking conventional reproduction involves a wide range of operations, all producing a 'physical-type' product which is passed on to the next stage for processing, converting and positioning into the required format.

Typesetting can be carried out on traditional phototypesetters which can only set type or other 'shapes' from negative master characters. Modern imagesetters, however, can set images of any form such as type, graphics or halftone. The output from the phototypesetter or imagesetter can be in paper bromide form or film (negative or positive). If in bromide form, the traditional pattern of *paper make-up* is followed, where the bromides are pasted down (waxed) to layout sheets, often pre-printed in non-photographic light blue for regular publications.

Photographs, transparencies, artwork, predominantly for black-and-white reproduction, are then mainly reproduced on a *graphics arts camera* using a range of contact screens for continuous tone work. Processed output from the camera can again be in paper bromide form or film, mainly negative: if the result is to be in positive form then it is just a matter of exposing or *contacting* the negative to another piece of light-sensitive material to produce the opposite effect, ie a positive.

Film make-up can either be negative- or positive-working. In the UK, negative film working is used mainly for single and spot colour work, with positive film working for four-colour process. For complex tint laying and image montaging, especially in four-colour process, the normal practice is to prepare *inter-negs* then contact to final positive film.

Colour scanners can either be operated as stand-alone units or linked to a workstation. As a stand-alone unit, it will involve a considerable amount of time and expertise in manual planning and assembly. If the scanner is linked to a workstation unit, it must have an electronic link-up.

Monochrome scanners are becoming popular for high quality resolution pictures as well as for high volume requirements. Cheaper flatbed and drum scanners are used, again with an electronic link-up.

Film make-up and assembly

Manual film planning, especially colour work, is an exacting, time consuming and expensive operation which relies heavily on the skill of the individual. If several cut-out illustrations, mask cutting and tint laying are included in a complex four-colour process job, then the operator may have to plan up to 20 or more negatives (inter-negs) for each set of final, positive, four-colour films. It is for this reason, ie the high cost and skill level required with conventional planning, that electronic methods of planning are becoming more popular.

Electronic reproduction

The principle of electronic reproduction is radically different from conventional reproduction in that the system creates and manipulates digital/electronic data. It only produces 'physical' or 'mechanical' data - such as completed finished flat artwork from a laser printer or imagesetter, also finished film or plates from a scanner or imagesetter - when required, while still retaining its digital form in the host storage system.

Typesetting/capturing of keystrokes ranges from *word processors, PC disks* or input from *OCR systems* into *electronic text files.*

Photographs, transparencies, artwork are scanned into the system, using a scanner either for colour or black-and-white reproduction. Digital cameras, as previously discussed, can supply digital images direct to the receiving DTP system. Graphics generation can also be produced using specialist software manipulation programs such as:

Adobe Illustrator - inputs scanned images or generates new images from arts package tools. It is capable of producing halftone shades as well as supporting colour process.

Aldus Freehand - another very powerful and precise graphics generating tool which allows the importation of images for manipulation and outputting as required; again supports colour processes.

Adobe Photoshop - one of the most popular graphic arts software programs, it has an image processing program for the creation and reproduction of colour and black-and-white images.

QuarkXPress - with its facility for four-colour separation and excellent typographic standards this program has become the 'standard' for many in the industry and, in fact, several suppliers such as Scitex and Crosfield have linked it into their own electronic systems.

Manipulation of images

Ease of manipulation of images is an undoubted strength of the electronic route where various WP/text capture, page make-up and graphic generation software programs can integrate to produce whatever result is required, dependent on the system installed.

Visuals and/or artwork can be reproduced from a laser printer or imagesetter; also visuals can be produced using expensive high-quality 'target' proofs such as Signature, Digital Matchprint, Iris ink jet or the cheaper desktop systems using thermal transfer, dye sublimation or ink jet.

Following revision of artwork into its final finished form, the images can be output onto *film* or *direct-to-plate material* by linking into a workstation unit with scanner or imagesetter as an output unit.

Colour reproduction using DTP links is only now beginning to approach a level considered to be of a reasonable commercial standard. Initially, at least, it was lacking in quality, often suffering from moiré or screen clash pattern, as well as being extremely slow in outputting.

One of the main drawbacks to DTP colour is the size of file required to process even the simplest colour work. To help overcome this, the capacity of the PCs used has been increased considerably.

Desktop colour systems have developed as a cheaper alternative to the larger colour EPC *(Electronic Page Composition)* systems. The quality of colour reproduction from a desktop scanner linked to a DTP make-up system tends to be inferior and much slower compared to that achieved by using the conventional colour drum-scanner or high-quality larger flatbed colour scanner.

To ensure higher quality colour output, systems have been developed which store the DTP-produced data in PostScript on a floppy disk and, through this translating medium, access conventional four-colour scanner systems. Two examples which adopt this approach are Scitex Visionary and Crosfield Studiolink.

Such systems form the link between DTP and EPC systems and are basically a software colour layout control program based on Mac or similar PostScript-driven PCs. Each system can prepare masks and colour tints; text capture is mainly through disks processed on a word processor or other data capture device such as an OCR scanner.

Initially, the transparencies are scanned conventionally using a colour scanner; PC viewfiles are also produced from the scan data. The viewfiles are entered into the PC system where text and graphics are prepared in an approved page format. This formatted disk is then fed into the relevant system which replaces the low-resolution viewfiles with the original high-quality colour scans. The composite page is then output on the scanner unit.

The colour sequence can be reversed so that the low-resolution scans used initially on the DTP system for page make-up are imported into the workstation where high quality resolution scans replace the DTP-created pictures.

The term DAR *(Digital Artwork and Reproduction)* is another addition to the growing modern pre-press glossary. It is basically another expression for EPC or DTC *(Desktop colour)* using DTP and digital forms of reproduction.

Electronic page composition/system workstations

During page make-up operations, all the subjects needed for a job - type, photographs, graphics and tints etc - are scanned or input into the system by other means. Also, as far as possible, the colour balance and other reproduction parameters are set up at the input stage.

A graphics display terminal allows the operator to view each colour original and if necessary to adjust the tonal range from highlight to shadow for each separate process colour. Retouching facilities include sophisticated 'airbrushing' where a flaw or mistake can be corrected and details added or removed from a picture without the need for re-scanning. Other features include taking items from one picture and incorporating them in another, the swapping of colours and pixel retouching. This involves enlarging a subject on the screen until the smallest area or electronic component can be seen. The operator can then move these components about so that any blemish can be removed or a detail swapped - for example, changing a dark-blue suit into a light-grey one.

Once all the page elements are approved, the operator can manipulate them on the screen in the correct position and rotate or vignette them according to the layout. Tints and borders are viewed in full colour and all systems offer the ability to incorporate text and graphics.

The final stage is to output the page or pages to a colour scanner or imagesetter. Ideally, the output is a single composite piece of film for each process colour separation. As a result of their use, EPC systems cut down considerably on film usage and their operation is particularly cost-effective on complex, highly-illustrated colour work - for example, holiday brochures, catalogues, colour supplements and mail order.

One of the major problems facing the DTC-driven users is speed of response and insufficient storage space, ie *data management*. Even a limited use of colour work creates huge amounts of PostScript data which has to be managed and manipulated, with the result that often operators spend a considerable amount of their time waiting for terminals to clear and RIPs to finish previous jobs.

Early forms of workstations used for colour reproduction had up to 200 Mb of storage, this has now increased up to 1.4 Gb and even 2.8 Gb. Recent developments in this area are the optical disk library jukeboxes, consisting of a battery of 32 optical disk drives providing an on-line capacity of about 20Gb - no doubt this will be increased even further in the future.

CD *(Compact Disc)* is a developing technology which offers tremendous potential in low-cost storage, as well as simple and effective transference of digital data.

Planning and assembly - punch-pin register systems

The use of punch-pin register systems is very important in assisting precision printing. This is especially true in conventional mechanical manual planning with complex, highly-illustrated colour work, and also with multiple-image exposure work.

It is essential that all printed images and printing plates, when positioned on the printing machine, are in the correct, pre-determined position. Some printers only use a punch-pin register system to link the layout sheet and montage foils to the plates rather than integrating it at every stage of production. The drilled or punched materials are registered to each other by alignment to the holes and slots created by the punch register system and held in position by the required pins.

Electronic imposition/automated planning and assembly

Electronic imposition is now established with the products and systems necessary for its success now in place, such as large-format imagesetters and appropriate software programs compatible with a wide range of popular DTP programs.

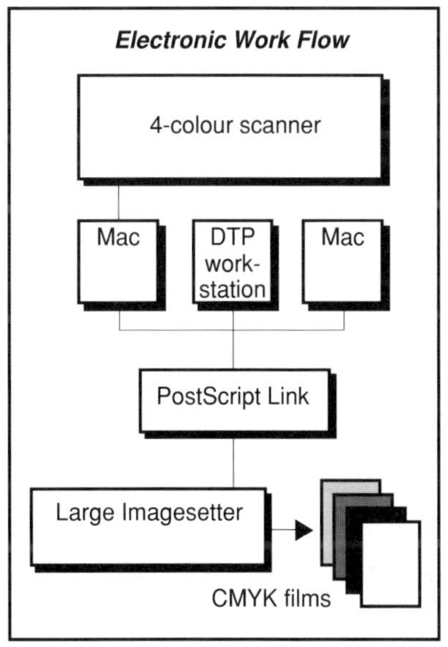

Large imagesetters are now available which can output up to eight A4 four-colour process pages in exact register on four pieces of final planned punch register film.

In the drive towards streamlining the planning processes, many companies producing large-format imagesetters are developing their own imposition software as well as linking in with other systems.

Figure 7.3: Diagram of an electronic work flow

Specialist DTP electronic page-imposition software programs include *Aldus Presswise, Ultimate Impostrip* and *Imposition Publisher*.

Aldus Presswise converts the pages of a *PostScript* file created from, for example, PageMaker or XPress into imposed printing form. It has the facility to merge pages from up to 32 PostScript files at any one time into standard imposition templates, utilising numerical controls to move page images to compensate for different binding margin requirements - see *Glossary* at the back of the book for a brief explanation of PostScript.

Ultimate Impostrip is available as a stand-alone package or, increasingly, built-in to imagesetter systems; it is claimed to give a cost saving of approximately 60% over conventional film planning.

Imposition Publisher is compatible with over 20 DTP packages, with access to all the recognised BPIF folding impositions, as well as the facility to rotate films in order to fit the output device medium, minimising bromide paper and film waste.

The choice between manual and electronic planning is not likely to be an absolute in either case for some time as often the printer is supplied with part film for a publication, or flat artwork which will be more cost-effective to plan using conventional planning techniques. In addition it is not always practical to wait until all images/data are available and complete before outputting in composite form.

Proofing

Proofs have two major purposes:

1) *for submission to the customer for checking and approval*
2) *to act as a working guide for the printer.*

There are many types of proofs which can be produced at the various stages of a printed job, each often serving a different purpose.

Monochrome/spot colour proofs such as galley, page, laser and ozalid etc, are all produced for checking the accuracy and position of the printed elements on the page.

Colour proofs such as the 'wet' and 'dry' varieties are mainly for checking the quality standards, colour registration and colour balance/ reproduction as compared with the original(s).

'Wet' colour proofs are so called because they use wet printing ink and are normally produced on specialist flat-bed proofing presses - the best known type being the *progressive proof* which shows each colour on its own and then superimposed in progressive sequence colour by colour.

'Dry' colour proofs cover a very wide range including *photo-mechanical, electrostatic, digital, non-impact, soft proofs* and *overlay film proofs*.The most popular and best known of these is the *photomechanical* represented by Cromalin, Matchprint and Agfaproof - the photo-mechanical is the proof most often requested by design agencies.

Presentation of proofs

Proofs can also be supplied to customers in different forms, eg:

single proofs - of one image or page;

backed-up proofs - perfected proofs, often on job stock;

imposed page proofs - showing laydown of pages in correct position as printed;

scatter proofs - images scattered over one sheet not in exact position;

progressive proofs - printed colours in sequence, individually and in combination;

production-proof - ultimate proof in that it is produced on actual production press and stock. *No other proof can match this for accuracy and closeness to the final production run.*

Classification of proofs

In recent years there has been a move to clarify the terminology and understanding of the purpose of different types of proofs. Four major areas have been identified:

visual proof - such as DTP thermal, laser or ink-jet proof, produced by a graphic designer or client, to indicate the overall concept and design required in the finished product.

typographic proof - such as ozalid, dylux, laser proof or bromide, produced to enable the client to check the typesetting, position of graphics and imposition of pages before the platemaking stage.

target proof - such as Cromalin, Matchprint and Agfaproof, or wet proofs, which are produced to represent as closely as possible the expected printed result, conveying press characteristics such as dot gain and colour accuracy.

contract proof - such as *pass on press* pull or approved target proof, which is 'signed off' by the customer for the machine printer to match and use as the *master pass sheet*.

It is important that the printer and customer both appreciate the value of the different stages of proofs so that they use each to their best advantage in picking up, for example, typographic errors, illustrations wrongly positioned or colour faults.

Printing surface preparation/ platemaking

Each printing process uses a different printing surface to suit its particular requirements.

Offset lithography requires a planographic flat printing plate with ink-accepting image areas and water-accepting non-image areas.

Lithographic printing plates may be made from metal, plastic or paper, which are treated to receive or retain properties which are conducive to the lithographic process. One of the main properties is a photo-sensitive surface coating, although some paper and plastic plates may be typed to receive the printing image. Lithographic plates are available as negative-working or positive-working.

Negative-working plates are processed with a photo-sensitive emulsion which *hardens* when exposed to light; conversely, *positive-working plates* are processed with a photo-sensitive emulsion which *softens* when exposed to light. Conventionally, both types of plates are prepared by placing the pre-sensitised unexposed plate in a printing-down frame with the planned foil consisting of montaged negatives or positives in a pre-determined position on top. The plate is then exposed to a high energy light source and processed manually or on an automatic plate processor.

Flexography and **letterpress** require a relief printing plate or block with raised image areas which are inked and non-image areas which do not receive ink.

Letterpress requires a relief block or plate which, in the case of flat-bed printing, must be made up to a type height of 23.317mm.

Printing surfaces can either be *originals* which means blocks, normally made from zinc or copper being made photo-sensitive, exposed to a negative, and then the non-image areas etched away with acid. Alternatively, the relief image is created by an engraving process.

Duplicate plates are copies or duplicates taken from the master mould or original plate. This type tends to be made of plastic or rubber produced from a moulding process and is mostly used in rotary letterpress printing where the plastic or rubber plates are mounted onto clear plastic sheeting as a printing sleeve to be positioned on the machine.

Photopolymer plates are presently most popular for letterpress and flexography. The plates are made from a photosensitive plastic which hardens under the action of ultra-violet light. Photopolymer plates are mounted on the rotary presses by attaching them to plastic sheeting sleeves, or alternatively they are exposed as 'one-piece' plates.

Apart from photopolymer, **flexography** uses *rubber moulded* or *hand-cut rubber stereos* as plates where the excess or non-image rubber is cut away with a sharp cutting edge. *Laser-engraved rubber plates* are used when high quality and fine definition flexo plates are required. The plate is processed by a computer-controlled laser selectively removing the non-image areas from the rubber-covered roller. It is then secured onto the press in position.

Gravure requires intaglio cylinders or plate with recessed image areas which retain ink and flat surface non-image areas from which the ink has been wiped.

There are two distinct methods of producing intaglio gravure cylinders - *etching* and *engraving*.

Etching a gravure cylinder is a complex and long process which requires sensitising a carbon tissue, on which is exposed firstly an overall grid screen, and then a continuous tone positive. The processed carbon tissue is then attached to the copper cylinder where the soft gelatine of the carbon tissue is dissolved away leaving varying degrees of hard, soft and little or no gelatine. When the cylinder is exposed to acid the unprotected areas of the cylinder will etch away to produce the solid areas, with mid-tone and highlight areas being formed, depending on the amount of protection given by the 'semi-hardened' carbon tissue.

Electronic engraving is tending to replace etching as the main means of producing gravure cylinders. It is a much simpler process where an engraving stylus cuts or 'digs' out the recessed cells, producing an intaglio cylinder or recessed image areas.

Screen requires a stencil processed frame or cylinder with clear, unblocked image areas through which the ink passes and non-image areas which are filled-in or blocked off.

Screen stencils are made from meshes of nylon, polyester or stainless steel as these have been found to be a better carrying medium than the original material, silk, from which the process first took its name, ie silk screen printing.

The process operates on the basis of the screen mesh being blocked off in the form of stencils to produce the non-image areas.

Stencils can be produced by *hand* and *electronic cutting*; as well as *photomechanical* where a light-sensitive material is exposed to a positive image. The normal image areas are hardened with light, leaving the unexposed image areas to wash away, creating a stencil area through which the ink will pass.

Direct-to-plate systems

Several different direct-to-plate systems are available, all having the common ability to produce printing plates without the intermediate use of negative or positive film.

There are two main types of direct-to-plate systems:

1) *direct from artwork to plate* using a camera or platemaker system

2) *direct from EPCS or laser imagesetter to plate*, often referred to as a *computer-to-plate* system.

In the case of making offset litho plates direct from artwork, a specialist camera with a reversing mirror mechanism is used to obtain a right-reading plate or, alternatively, a specialist platemaker unit is used, both often with built-in processing facilities. The plate material can be metal, using the electrostatic process with liquid or powder toner to produce a plate suitable for newspapers and bookwork, for example.

Alternatively, the plate material can be silver-coated polyester/ silvermaster-type material. The cheapest type of plate is, in fact, the paper plate or master, where the specially-coated paper sheet receives the image areas by being passed through a photocopier, laser printer, or by directly typing onto the paper plate.

The computer/imagesetter direct to digital platemaking route is an area with great potential for fast processing of quality, cost-effective plates; currently the main material being used is *silvermaster-type polyester*, with several companies working on producing aluminium plate materials or alternative film-based plates.

For many years there has been a considerable amount of research and development into finding a 'waterless' plate which will operate under normal printing conditions - the *Toray plate* has been the established market leader in this area for some time.

Computer-to-press developments will no doubt advance in the future, but at the time of going to press this area was still relatively new for the main established printing processes; new digital printing processes are, however, becoming established in this area.

8 Methods of working and impositions

Each printing company uses a range of printing equipment which has been installed in response to either identified market needs, production and/or marketing expertise in a particular area, or as a service in some form or other to a larger organisation. The main objective, whatever market sectors a company operates within, is to use the equipment and other resources at its disposal as efficiently as possible - in terms of production control this is covered in Chapter 6.

It is essential that technical, customer service and administrative staff, in providing an efficient and responsive customer contact, have an understanding of the areas to be considered when planning jobs to obtain the most cost-effective and practical use of the equipment available.

This chapter concentrates specifically on the areas of methods of working and impositions which to a large extent are dictated by the size and type of printing machines and finishing facilities available. Subject areas covered are: sheet-fed presses - multiple-image printing, half-sheet work, work-and-turn, work-and-tumble, sheet work, printed first-and-third, work-and-twist; choosing the method of working; flat plans; insetted and gathered forms of binding sections; full-width and narrow-width web-fed presses.

Methods of working considerations

When deciding on the preferred method of working for any job, several factors need to be considered:

a) *the quantity to be printed* - leading to considerations of multiple-image printing (where appropriate) on long runs.

b) *colour 'fall' of printed pages* - for example, whether a booklet or brochure is printed in four-colour process throughout, or whether the publication is a mixture of monochrome and four-colour process pages so that grouping of common colour pages can be used cost-effectively.

c) *printing press sizes and number of units available* - for example, a possible selection of small offset, B2 or B1 sheet-fed presses in one, two, four and five colours. Alternatively, heat-set web offset presses in eight-, 16- and 32-page configurations in four, five and eight units.

d) *types of printing press to be used* - for example, printing one side only, perfector or convertible.

e) *method of binding* - insetted sections as used in saddle-stitching have half the pages (lower half) on one side of the centre spread, and half the pages (higher half) on the other side. Gathered sections, for example, as used in thread sewing and perfect binding, break the pagination down into sections of consecutive numbering, (see page 131).

The finishing operations which have greatest impact on the method of working and hence imposition are *cutting, folding* and *binding,* (see *Chapter 11).* The laydown of the pages on the printed sheet or web must be capable of being cut or slit and subsequently folded into the correct sequence by the equipment available. In the case of reel-fed label printing the main considerations are the size of available cylinders, width and circumference, also the related finishing equipment such as flat-bed or rotary cutters.

Imposition

Imposition is the arrangement and assembly of printed images into a pre-determined format so that, when bound, each printed page will appear in the correct sequence and position. The approach to imposition applies equally to all the major printing processes, apart from recognising the fact that offset litho requires a right-reading offset litho plate, whereas conventional letterpress, flexography and gravure require wrong-reading printing plates or cylinders. Screen printing operates from a right-reading stencil mesh, emulsion-side down.

All imposition schemes work backwards from the finished printed product. As there are considerable practical differences between sheet and web printing they are dealt with separately in this chapter.

One of the first considerations when planning how a job is to be printed is to ensure that the available finishing equipment can complete the work in the desired way, because once a job is printed, a mistake in the imposition is difficult, if not impossible, to correct. When in doubt about any job, the person responsible for initiating the work should always consult the finishing department, especially if an outside supplier is to be involved. In particular, such details as the position of grips and trims should be checked. When an imposition scheme has been supplied by a customer it should be checked with the finishing department unless it is a scheme with which all are thoroughly familiar.

Sheet-fed presses

There are various methods by which paper/board may be printed, or worked, on sheet-fed printing machines and these depend on:

a) whether the sheet requires to be printed on one side only, or whether it is to be perfected or backed (also sometimes referred to as duplexed) - that is, printed on both sides; and

b) the imposition scheme - the arrangement of the pages so that they appear in correct sequence when the sheet is printed and folded.

Multiple-image printing

Work which is not to be bound in any way, such as a leaflet, carton or label, and printed on one or both sides, presents few complications in terms of how the laydown of images is decided. Large print quantities, to be most effective, will normally result in *multiple-image printing*.

Figure 8.1 shows a printed image which is to be printed 10-up: if the quantity required is one million, this will result in a print run of 100 000, times the number of required press passes to complete the job.

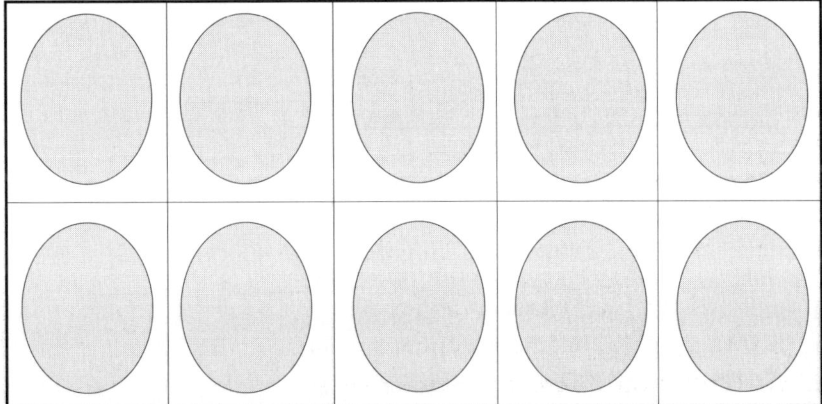

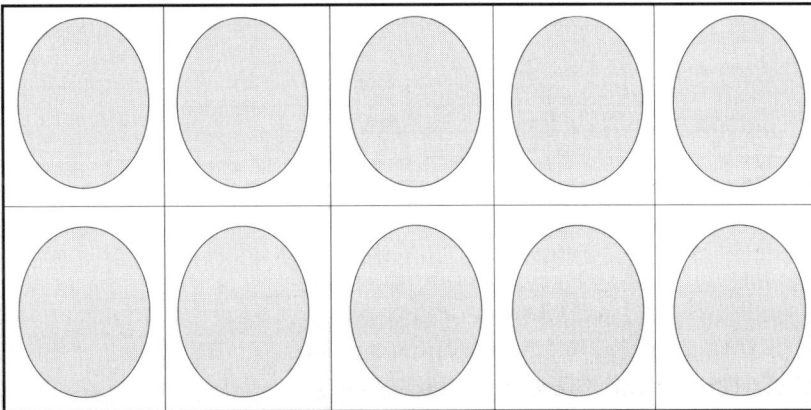

Figure 8.1: 10-up multiple-image laydown

When paper/board is required to be printed on both sides and bound, different impositions will be used for printing the sheet depending on whether the imposition is for half-sheet work or sheet work. These terms are applicable to any size imposition of two pages or more, and the methods of working are explained on the following pages.

Half-sheet work

In the method of working known as *half-sheet work*, all the pages in a section are arranged in one imposition. By printing one side of the sheet from the imposition and then turning it over at the end of the run to print the other side, two identical copies are made from one sheet. The sheet is then cut in half, each half of the sheet having all the pages in the imposition printed on it. Half-sheet work may be either *work-and-turn* or *work-and-tumble,* depending on whether the sheet is *turned* or *tumbled* prior to perfecting. These impositions are naturally suitable only for single-side press working, rather than perfecting on perfector or convertible presses. When reference is made simply to 'half-sheet work', work-and-turn is normally implied.

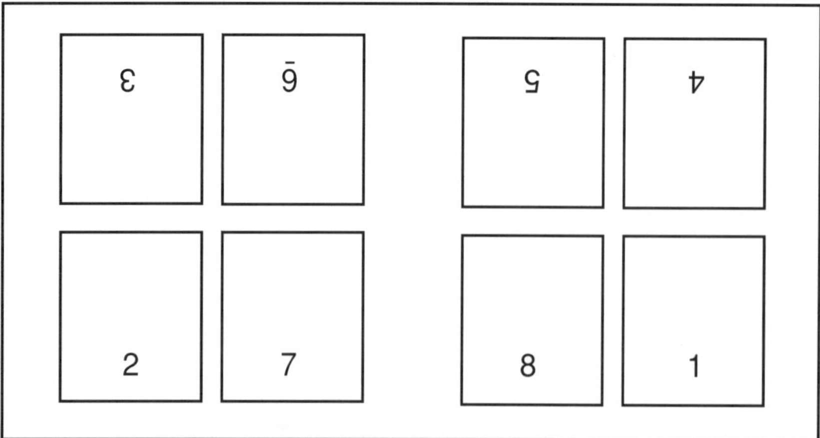

Figure 8.2: Eight-page work-and-turn imposition

With half-sheet work, only one imposition is used and there is therefore only one make-ready, but the size of the plate and the working size of the sheet and press have to be large enough to accommodate all the pages that are to appear on both sides of the sheet. For example, a sheet to be printed eight pages on one side with eight pages on the other would be arranged as a sixteen page imposition and the working size of the sheet would be twice as large as the complete sheet required.

Thus half-sheet work, single-set working, requires half the number of working sheets as there are impressions because each sheet, when cut in half, produces two complete copies of the job - for example, 30 000 copies printed half-sheet work of a sixteen-pages A4 portrait booklet would require 15 000 sheets plus overs of sheet size SRA1 (640 x 900mm). Paper printed half-sheet work has to be cut or slit in two before folding.

Work-and-turn

In this method, the sheets are *turned* on the axis of the short side, which ensures that the long edge of the sheet, gripped when printing the first side, is also presented to the grippers for the second printing, (see *Figure 8.2*).

In this type of imposition, the plate or set of plates backs itself. In order to ensure correct register of pages during backing-up, the side lay is changed from left to right and, in this way, the same edges of the sheet are fed to the lays each time the paper passes through the machine.

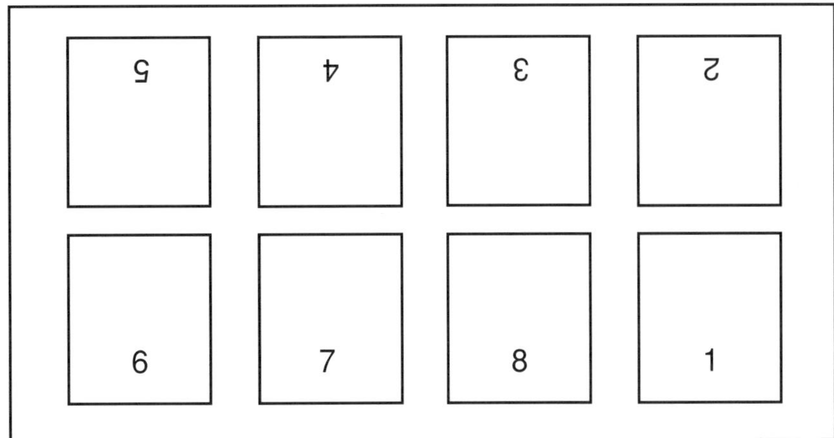

Figure 8.3: Eight-page work-and-tumble imposition (roll folded)

Work-and-tumble

As with work-and-turn, both the inner and outer pages of the sheet are imposed together. All the paper is printed on one side and then tumbled from front to back before the reverse side is printed from the same imposition (see *Figure 8.3*).

Unlike the procedure in work-and-turn, the sheet is turned on the axis of its long side - that is, *tumbled* - where the two long edges are fed to the grippers in turn. For this reason adequate gripper margin has to be allowed on both long edges and the paper has to be perfectly square before printing in order to ensure register of pages in the back-up. Thus a tumbled sheet produces two complete printed copies, as does a turned sheet, the size and number of working sheets being the same for both methods.

Work-and-tumble is frequently used for six- and 12-page jobs and other instances such as an eight-page roll-fold, which would tend to result in an elongated or awkward-shaped sheet if printed by work-and-turn.

Figure 8.3 is an example of an eight-page work-and-tumble imposition suited to roll-fold finishing.

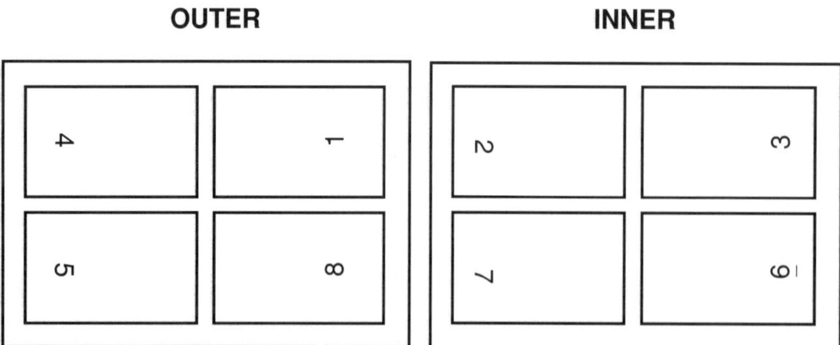

Figure 8.4: Eight-page sheet work imposition (inner and outer)

Sheet work

In full sheet work, the pages to be printed on each side of the sheet are arranged in two separate impositions, *inner* and *outer*. The outer imposition always contains the first and last pages of the section and prints one side of the sheet, while the inner imposition always contains the second page of the section and prints the other side. Whereas half-sheet work produces two copies of a section from one imposition printed twice, sheet work produces one copy printing two sets of plates (inner and outer) once each as shown in *Figure 8.4.*

Paper is printed on one side from one imposition and is then printed from the other imposition on the reverse side, the same edge of the paper being fed to the grippers for each working. Thus, two sets of plates are needed to complete the sheet and a separate make-ready is required for each. As each imposition contains the pages for one side only, the working size of sheet is the same as that required for folding. In this way each perfected sheet produces one copy and therefore the same number of sheets as completed copies are required.

Apart from some small offset machines, the longer dimension of the sheet is fed to the front lays in printing - this ensures better register and greater accuracy, because the effect is proportionately less than if the shorter dimension is used, as well as there being less impact due to paper/board stretch or shrinkage.

For jobs consisting of several sections, such as bookwork, sheet work often produces economies in finishing, by eliminating the need for guillotining after printing, which is required with half-sheet work. Long run, multiple-section jobs made up of sections of the same number of pages are generally printed sheet work, the sections sometimes being produced on pairs of machines.

Two less-commonly used methods of working are printed *first-and-third* and *work-and-twist.*

Printed first-and-third. Sheets are printed in such a way that, when folded, the printed matter only appears on pages 1 and 3. Sheets for duplicate pads with the first leaf printed and the second leaf plain are often printed in this way in order that they may be folded in sections for binding.

The cheapest range of greetings cards are also often printed by the first-and-third method.

Work-and-twist. Jobs which are printed twice from the same printing image on the same side of the sheet but from opposite ends. An example of its use would be with a sheet printed one side only with rule work as cross rules in one half of the sheet and down rules in the other half. After the first printing, the paper is twisted through 180° so that the former left-hand side is on the right-hand side and then printed on the same side from the same printing image.

This method of working is seldom used nowadays - its popularity waned due to the decline of hot-metal letterpress and its replacement by offset litho, assisted by the introduction of simplified rulework available with photocomposition and imagesetting.

The BPIF's publication *Standard Folding Impositions* contains 27 different impositions, in diagrammatic form with text commentary, varying from an eight-page section to 4 x 16-page sections.

The booklet sets out imposition schemes suitable for most all-buckle and combination folders, plus some for all-knife machines.

Choosing the method of working

Location of colour pages and a steady flow of printed sheets to the finishing department usually govern the method of working. Where colour and time considerations are not important, half-sheet work can be used as economically as sheet work methods. Sheet work helps to overcome the problem of having to keep a machine idle while waiting for the ink to dry on the first side before printing on the other. This is because the first imposition (say outer) can be lifted as soon as the first side has been printed and the reverse side (say inner) need not be put on until the sheet is in a fit condition for backing-up. In half-sheet work the machine may have to be kept standing before the sheet can be backed-up from the same plate, especially on short runs and also where heavy solids have to be run.

41	42	43	44
45	46	47	48
49	50	51	52
53	54	55	56

Figure 8.5: Flat plan covering colour fall of 16-page section

Flat plans

When producing periodical publications such as magazines, journals and other structured format work, it is common practice for the publisher/customer to supply the printer with a *flat plan* indicating the required pagination for editorial and advertising pages, or other matter; on the other hand it can be used simply to indicate the sequence of copy or colour to be used on a publication such as an annual report.

As an alternative a *production dummy* may be supplied as a made-up copy or series of folded sections. The dummy will normally be produced from plain paper folded to the finished bound size, with details marked-up of page numbers, copy, illustrations, numbers of colours etc. The publisher, designer or customer has the task of juggling and manipulating the overall contents into a coherent and cost-effective publication. The printing of most periodical publications are produced under contract where the printing company quotes various options of colour usage within certain cost parameters - for example, the contract for a monthly magazine of 96 pages plus cover may state that the publication is saddle stitched and that the text is to be printed in 12 x 8-page sections as 6 x 8 pages in four-colour process and 6 x 8 pages in black only, split evenly throughout the publication.

Figure 8.5 illustrates a flat plan for the centre 16 pages of the text, ie pages 41 to 56 - the shaded areas indicate the pages planned in four-colour process, blank areas indicating the pages planned in black only.

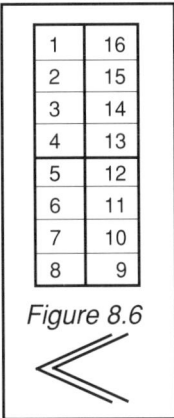

1	16
2	15
3	14
4	13
5	12
6	11
7	10
8	9

Figure 8.6

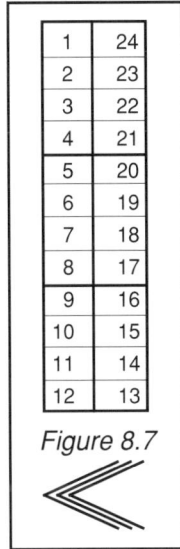

1	24
2	23
3	22
4	21
5	20
6	19
7	18
8	17
9	16
10	15
11	14
12	13

Figure 8.7

1	32
2	31
3	30
4	29
5	28
6	27
7	26
8	25
9	24
10	23
11	22
12	21
13	20
14	19
15	18
16	17

Figure 8.8

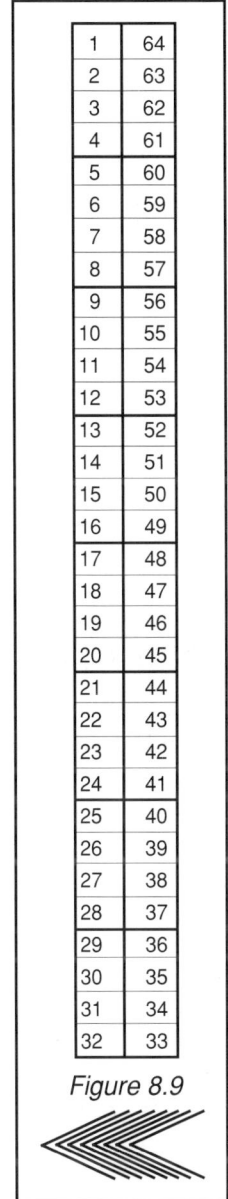

1	64
2	63
3	62
4	61
5	60
6	59
7	58
8	57
9	56
10	55
11	54
12	53
13	52
14	51
15	50
16	49
17	48
18	47
19	46
20	45
21	44
22	43
23	42
24	41
25	40
26	39
27	38
28	37
29	36
30	35
31	34
32	33

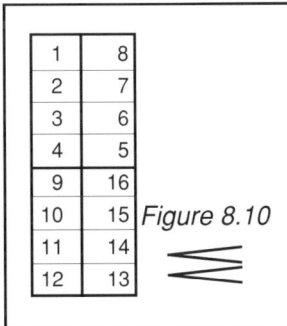

Figure 8.9

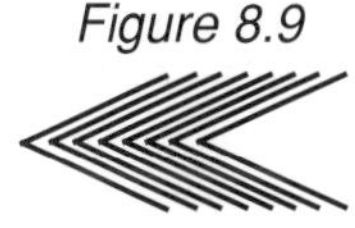

1	8
2	7
3	6
4	5
9	16
10	15
11	14
12	13

Figure 8.10

1	8
2	7
3	6
4	5

9	16
10	15
11	14
12	13
17	24
18	23
19	22
20	21

Figure 8.11

1	8
2	7
3	6
4	5
9	16
10	15
11	14
12	13

17	24
18	23
19	22
20	21
25	32
26	31
27	30
28	29

Figure 8.12

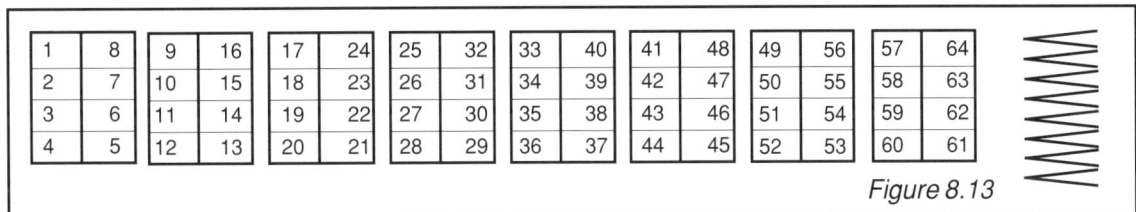

1	8	9	16	17	24	25	32	33	40	41	48	49	56	57	64
2	7	10	15	18	23	26	31	34	39	42	47	50	55	58	63
3	6	11	14	19	22	27	30	35	38	43	46	51	54	59	62
4	5	12	13	20	21	28	29	36	37	44	45	52	53	60	61

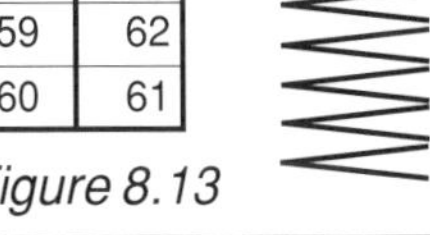

Figure 8.13

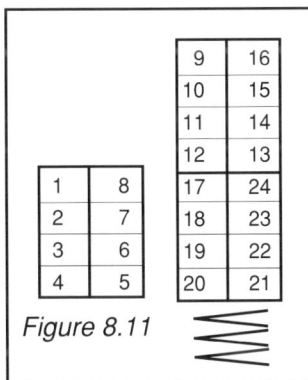

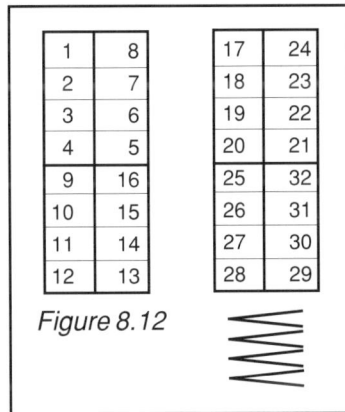

A careful study of the flat plan will show that the outer section pages are planned to be in four-colour process with the inner section in black.

Any deviations from the specification, including the use of colour outside the agreed flat plan, will incur extra costs. The flat plan is used by the printer as a guide to the colour fall of pages, which has a major impact on deciding the method of working.

Insetted and gathered forms of binding sections

Flat plans or production dummies need to take into account the method of binding for a publication. For periodical-type work such as magazines, journals, catalogues and more general-type work such as brochures, parts lists and manuals etc. there is the option of using *insetted sections for saddle-stitched binding* or *gathered sections for mainly adhesive or thread-sewn binding* (see page 131). In bookwork-type work where pagination is normally well over a hundred pages on bulky stock, gathered sections is the chosen method of binding; publications thicker than 7mm are unlikely to be suitable for insetted, saddle stitched work. The method of calculating the pagination (page numbers) in the two types of bound sections is quite different.

Insetted sections

Figures 8.6 to 8.9 illustrate four examples of insetted bound work covering 16-, 24-, 32- and 64-page publications split up into 8-page sections as follows:

1-4 and 13-16, 5-12;

1-4 and 21-24, 5-8 and 17-20, 9-16;

1-4 and 29-32, 5-8 and 25-28, 9-12 and 21-24, 13-20;

1-4 and 61-64, 5-8 and 57-60, 9-12 and 53-56, 13-16 and 49-52, 17-20 and
45-48, 21-24 and 41-44, 25-28 and 37-40, 29-36.

The main rules of insetted imposition (where right-angled folds are adopted) can be established from these examples, these are:

1) the sequence of pages run down to the *'half-way figure'* and back again.

2) pairs of pages add up to one more than the total number of pages.

3) on each section the pages are linked across the fold so that, for example, the first 4, 8 or 16 pages etc are 'paired' with the last 4, 8 or 16 pages respectively.

Gathered sections

Figures 8.10 to 8.13 illustrate four examples of gathered bound work - note the imposition rules as established for insetted work do not apply to gathered work. It can be seen from the figures that the pagination of gathered sections is much easier to establish - 1-8, 9-16, 17-24, 25-32, 33-40, 41-48, 49-56 and 57-64 respectively.

Full-width web-fed presses

All conventional web-fed printing for periodicals and newspapers is printed on perfectors where the method of working and imposition terminology is fundamentally different to sheet-fed printing.

Web perfectors always operate on a similar basis to sheet work, although instead of inner and outer impositions, web printing relates to *top side of web* and *bottom side of web.*

Other operations which are peculiar to web-fed printing is section/ signature delivery as *collect* or *non-collect*. Collect is where the finished folded product is insetted into one set as it is delivered off the press; non-collect is where the delivery is split into two sets or separate folded sections.

Imposition for a web-fed press is governed largely by the folding equipment fitted at the end of the press. The main exceptions occur if a sheeter or a reel-up unit is fitted in place of a folder. In such cases a sheeter will cut the printed web into sheets which can be finished in the same way as sheet-fed jobs, while a reel-up unit will re-reel the web for further processing.

There are many differences in the workings of web folders compared to sheet folders. In particular, while only one sheet of one section can be folded at a time in conventional sheet work, several webs (which may be of varying widths) containing different sets of pages can pass through the same web folder at the same time. Thus for example, a five-unit web-offset press, capable of printing 16 pages size A4, five colours on each side of the web at once, could produce a collect 32-page section. This would be done by running one web through four perfecting units to produce 16 pages in four colours and running a second web through one unit to produce a further 16 pages in monochrome. It is normally possible to arrange for the colour or monochrome pages to fall at the outside or towards the centre of the section as desired.

Narrow-width web-fed presses

Narrow-width web-fed presses are available in a wide selection of formats capable of producing a very comprehensive range of printing and print-related products.

The two main areas for this type of machinery are *continuous stationery* type products and *self-adhesive label* type products.

Continuous stationery and business forms can be either single part or multi-part, reel-to-reel or reel-to-pack. The machines can either be fixed-cylinder size presses or variable-cylinder size presses.

On the larger size presses the most popular cylinder sizes are 24.75", 24", 22", 20" and 17" circumferences. There are also the less popular sizes of 28", 26" and 18". A 24" cylinder press would be able to produce forms in derivatives of 24" - that is to say 24", 12", 8", 6" and 4" deep. A 22" cylinder press could produce 22", 11", 7.33" and 5.5" deep forms.

The width of the cylinders is from around 20" up to 30" wide. *Figure 8.14* illustrates a 26" wide x 24.75" circumference cylinder printing 6-up of an A4 form 8.25" (210mm) x 11.75" (297mm), ie 2 x 11.75" across the cylinder and 3 x 8.25" around the cylinder.

On the smaller-size presses the cylinder widths vary from 16.5" (420mm) or below, to circumferences as small as 11" (279mm).

Narrow-width reel-fed self-adhesive type machines are normally available in up to 250mm wide. Labels are often delivered one wide-to-view on the web. If printed more than one wide, a slitter unit will cut the number of labels across down to one wide.

The number-up calculations for narrow-width reel or full-width web machines are made from the number out from both the width of the printing web being used and the relevant cut-off, or subdivisions of the cut-off, of the printing cylinder.

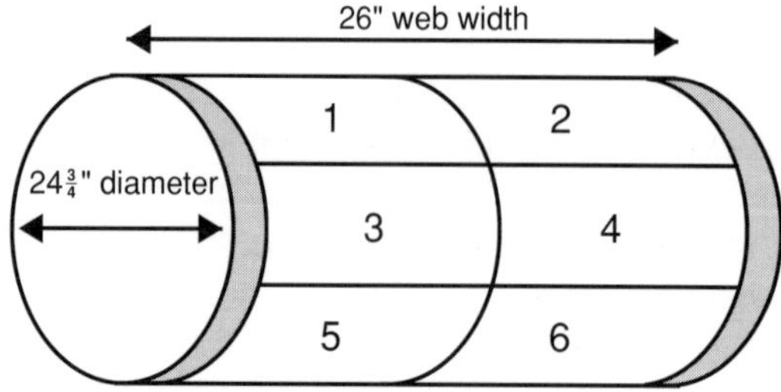

Figure 8.14: Printing cylinder illustrating 6-up A4 forms from one cylinder

9 Print specifications, reproduction standards and quality control

With the print-buying customer becoming more discerning and directly involved in the printing process chain through digital links such as ISDN *(Integrated Services Digital Network)* and DTP, it has become even more important for the customer and printing company to establish unambiguous and concise print specifications for each job and contract to ensure both are correctly interpreting the requirements of the other party.

This is essential at the present time where a multiplicity of conventional and electronic means of reproduction are used throughout the printing industry.

This chapter sets out to highlight the areas where the sales administrator/account executive needs to be aware of potential difficulties, such as in receiving and interpreting customer instructions, also in taking receipt of materials supplied by customers. Other areas requiring an increasing level of knowledge and awareness in the interface with customers is in quality control and reproduction standards.

Specific subject areas covered in this chapter are: print specifications - job descriptions and specifications, extra costs incurred by the printer and materials supplied by the customer; reproduction standards - preparation of camera-ready copy, line illustrations and halftone reproduction, designers' and printers' spreads; print specifications and standardisation - colour bars or strips and dot gain; quality control - quality control aids; colour management systems.

Print specifications

Printing, as with all other industries, has a certain modus operandi and to ensure the most efficient and cost effective interfacing between print buyer/customer and printing company all parties in the print chain need to state their requirements in a clear, concise and objective manner.

In ensuring the correct interpretation of the customer's requirements, the importance of the printing company preparing a tight and accurate job description and specification cannot be over-emphasised.

The consequences of not clearly establishing the customer's requirements at the beginning of an enquiry and/or order will lead to misunderstanding and misinformation between the printer and potential customer.

Additional 'extra' costs to the customer due to changes in the original specification can run into £00s if not £000s, eg DTP disks not formatted as agreed; re-origination of pages due to reductions or additions in the amount of copy or illustrations required; alterations to colour pictures after final colour proof stages and author's corrections, etc.

The more detailed and complete the job specification is, the more chance a printing company has of preparing an accurate and comprehensive quotation, which in turn considerably improves the opportunity of the job being completed to the customer's satisfaction, if the enquiry stages prove positive.

It is preferable for both parties if the printing company receives written instructions, either in the form of a letter from the customer or a completed enquiry form, along with, where possible, a dummy, layout, detailed flat plan or similar type of print sample.

The prepared enquiry with all other relevant details should be checked to ensure a full and clear understanding of the requirements. Irregularities or conflicting instructions should be noted and raised with the customer where appropriate. Customers welcome the opportunity to discuss and check their job specifications at an early stage of the enquiry if they realise it leads to a full and unambiguous understanding of the requirements on both sides.

The customer expects a firm, competitive quotation from a dependable printing company, in return the printer aims to produce a quality product yielding a satisfactory profit.

To assist customers in checking their print requirements and/or comparing quotations for the same job from different printers, it is essential that printers prepare comprehensive job descriptions and specifications which should include the information indicated below:

1) *Basic details of each job* - including number of copies required; size; number of pages; brief description of the job.

2) *Printing process(es) to be used* - eg sheet-fed or heat-set web offset litho.

3) *How many colours the job is printed in* - indicating clearly where colour falls on each page; also if the proposed job is, or is not, printed throughout in either one-, two- or four-colour process; additional 'special' colours or coating finishes, overall or spot as applicable, etc.

4) *Method of finishing* - including trimming, with or without bleed; folding scheme, especially if it includes the more unusual folds, such as gate, concertina, etc; folding to final size; saddle stitched 2 wires etc.

5) *Material specifications* - ideally type of paper and/or board; also the substance in g/m^2 or caliper for board weight.

6) *What is to be supplied by the customer* - in the way of artwork or illustrations, eg varying from complete camera-ready artwork being supplied to the other extreme of only rough sketches/visuals to be supplied, with the printer having to prepare all typographic and graphic reproduction.

7) *Number and type of proofs to be supplied* - eg photocopies, dry or wet colour proofs, etc.

8) *Schedule(s)* - clear indication of the date(s) on which 'copy' from the customer in the form of disk, camera-ready copy or film, etc is to be supplied; plus a list of proof out/proof returned dates, 'pass for press 'dates if appropriate, dates agreed for advance copies followed by completion of despatch.

9) *Packing and carriage details* - eg banded in 25s, cartoned in 500s. Delivery in bulk to one address.

10) *Payment terms* - eg nett monthly or 30 days from date of invoice.

Example of two job descriptions and specifications

'PRINTING MATTERS' magazine
30 000 copies, size 210 x 297mm, 96 pages text plus cover.
Cover printed sheet-fed offset litho and text heat-set web offset.

Printing company to plan and make plates from clean final film page positives supplied. Film to be one piece per page, right-reading emulsion-side down and reproduced to agreed specification.

Printed in four-colour process throughout with text on 70g/m^2 blade coated mechanical paper and cover on 130 g/m^2 gloss art paper.

Folded sections gathered, collated, perfect bound with cover drawn-on and trimmed flush to size. Shrink-wrapped in 50s, stacked on pallets and delivered to one address in central London.

'HOW TO BECOME A PRINTER' booklet
20 000 copies, size 148 x 210mm, 16 pages text plus cover.
Printed sheet-fed offset litho with cover printed green outside spread only, inside spread in black only, with text pages printed throughout in black only.

Cover scored and folded, text folded; cover and text insetted, saddle stitched two wires, trimmed to bleed.

Materials: Cover - 250g/m^2 one side cast coated board
Text - 110g/m^2 matt coated cartridge.

Cover to be supplied as one piece of finished line artwork, ie pre-screened where appropriate, and inside spread as one piece of line artwork with both pages planned together.

Text supplied as 12 separate complete made-up pages of photoset bromides with four s/s line illustrations.

Proofs: four sets photocopy proofs imposed in four pages, ie A3 size, to be supplied.

Packing/Carriage: Banded in 25s, cartoned in 250s. Delivery in bulk to one address in London.

Extra costs commonly incurred by the printer

There should always be a clear understanding of what potential extras may appear on the final invoice raised by the printing company for any job. This is often an area of contention between printer and customer since it normally constitutes an area outside the agreed price from the approved quotation, order or specification.

The most common extras charged include:

Poor quality originals - where copy supplied for typesetting is not clear and legible; thereby necessitating additional work and therefore extra costs.

Extra proofs - where proofs additional to those quoted for are requested by the customer.

Additional delivery - instructions are received which include expedited and increased delivery destinations.

Materials supplied by the customer

such as:

Paper and board - extra cost incurred to cover the shortfall of material supplied, sufficient to include satisfactory overs; also due to materials supplied being below required quality standard.

Film - supplied not as quoted, eg unplanned instead of planned and imposed on foils; also extra cost for contacting negatives when specification stated positives to be supplied.

Plates - supplied damaged and/or unsuitable for designated printing presses.

Author's corrections - corrections to original manuscript, disk supplied, or artwork; deletions or additions to plates at proof stages.

Additional services - postal charges; mailing; banding or making-up into sets.

Premium rates - overtime working necessitated by late delivery of copy, artwork, film or materials supplied.

Extra costs would normally only apply when the customer has deviated from the agreed specification or production schedule.

Reproduction standards

The preparation of camera-ready copy

The quality of artwork prepared or supplied to a printer has a considerable effect on the final printed result. Poorly prepared artwork consisting of a montage of separate items, such as a myriad of typeset bromide strips with illustrations and graphics prepared on thick board material runs the risk of bits falling off, resulting in uneven and misaligned printing.

Camera-ready copy is defined as copy that can be reproduced photographically without further preparation by the printer, except tint laying and the inclusion of any illustrations when supplied separately.

Artwork in the form of camera-ready copy should be mounted down firmly onto a flexible board base and protected with an overlay sheet, often a thin or transluscent material marked-up with the reproduction details - artwork, if possible, should never be rolled or bent.

Phototypesetting and imagesetting bromide material should be exposed and developed properly so as to ensure a high contrast between the dark image areas and the surrounding white non-image areas, which should be thoroughly washed and cleansed of any residual chemicals otherwise they will tend to yellow and discolour after some time has elapsed.

When minor corrections have to be made to the original artwork it is important to output a whole paragraph or section, rather than individual lines which are awkward to align in exact position. The stripped-in bromide paper edges are also easily picked up by the camera or scanner and have to be spotted-out before or during platemaking.

Pasting the artwork material on the base artwork layout sheets can be carried out by various methods such as using a spirit-based adhesive, a rubber-based adhesive or the more popular system of paraffin-based wax. The latter has the advantage of allowing considerable re-positioning until the correct placement has been obtained. Adhesive-based systems are more permanent, being difficult to reposition without causing damage to the base sheet.

If wax is used it is melted and applied by the roller of the waxing machine to the reverse side of the paper-based typematter and graphic elements to be pasted down. The wax-backed material is positioned initially by light pressure, then applied with firm pressure for final positioning.

The accurate cutting of any line or insertion is essential and is normally done by a sharp scalpel and straight edge or steel ruler. Ruled lines can either be drawn in with a drawing pen, applied in the form of pre-ruled adhesive tapes or produced direct on the imagesetter.

Line illustrations

Line illustrations can be introduced by pasting in the original providing it is to the correct same size as the rest of the artwork, and that it has been produced on a suitably-thin material for reproduction.

If there is a significant difference between the thickness of the illustration's base material and the typeset bromide material, problems will again be experienced due to unwanted edge marks.

When the original needs to be enlarged or reduced, a photographic line bromide to the required size will be produced. This is then cut to size, waxed and placed into position on the layout.

Halftone reproduction

The reproduction of continuous tone originals using paper make-up procedures falls into two separate methods.

The first method, using screened bromides, involves reproduction of the originals through a halftone screen onto photographic bromide paper.

Screened bromides are then planned into position on the base artwork with the typematter, producing a complete composite line negative - *'dot for dot'* in the screen areas.

This method is used extensively in newspapers, monochrome magazine work and other areas where a screen ruling of no finer than 48 lines per cm (120 lines per inch) is normally acceptable.

If a finer high-quality halftone reproduction is required, the system utilises separate *screened negatives*. Using this method, the exact spaces to be occupied by the continuous tone originals are blocked or masked out with black or ruby-red masking film cut to exact shape on the layout.

When completed, the assembled typematter and window masks on the layout base are photographed to produce a line negative. The continuous tone originals are then screened and reproduced as negative film, after which they are then positioned in the clear window mask areas.

Positioning the screened negatives can be done either by taping in the space provided; by *mortising* (that is, cutting out the transparent picture areas and inserting the screened negatives into position), or producing two separate foils, one for the linematter and one for the screened matter and printing down twice during platemaking.

Artwork preparation and checking

When camera-ready copy has been prepared it should be checked thoroughly by the customer, ensuring that all possible changes are incorporated before the printer is given the go-ahead to proceed to reproduction. It is always in the customer's and printing company's best interests to thoroughly discuss any job at the earliest opportunity - this certainly refers to instances where the customer prepares or commissions another party such as a graphic designer or design studio to produce camera-ready copy in hard copy artwork, disk or film form.

With the move towards electronic reproduction - see *Chapter 7, Creating printed images* - it is now becoming more common for 'artwork' to be output in one composite piece per colour, including *type and graphics combined.* Even further to this is the increasing adoption of camera-ready artwork prepared digitally in imposed form to be output as finished film or through a direct-to-plate system.

Often a printing company is provided with a variety of elements, with varying specifications for the same job which will add considerably to the overall cost: eg negative and positive film supplied for offset litho reproduction - unless a special processing unit such as the Horsell Graphic Industries *Gemini platemaking system* is available, then the negatives will need to be contacted to positives or vice versa as offset litho plates are either negative- or positive-working, not both.

Production and quality problems will arise if film is supplied wrong-reading emulsion-side down if the film is intended to be used directly for platemaking. Artwork and originals produced out of proportion, ie different sizes, also varying reductions and enlargements, will incur additional reproduction costs compared to the work being supplied same size or at least in groups of the same proportioning basis. Whenever possible the client, when commissioning or ordering finished artwork or film for the printer, should arrange for the work to be supplied in *printers' pairs* rather than designers' pairs.

An even better arrangement, if multiple pages such as four- or eight-pages are to be supplied, would be that they are supplied in line with the imposition agreed by the printer; also that the correct space is included in the backs (binding edge) suited to the method of binding and the position of any section in the bound publication.

Figure 9.1 illustrates pairs of pages appropriate to an eight-page saddle stitched booklet as *designers'* and *printers' pairs*. When designing and laying out pages, the designer will naturally lay pages out in pairs as they would appear in the final printed copy, ie appearing double page spreads; the printer, however, lays out pages to the required imposition so that when printed, folded and bound the pages will fall into the correct sequence.

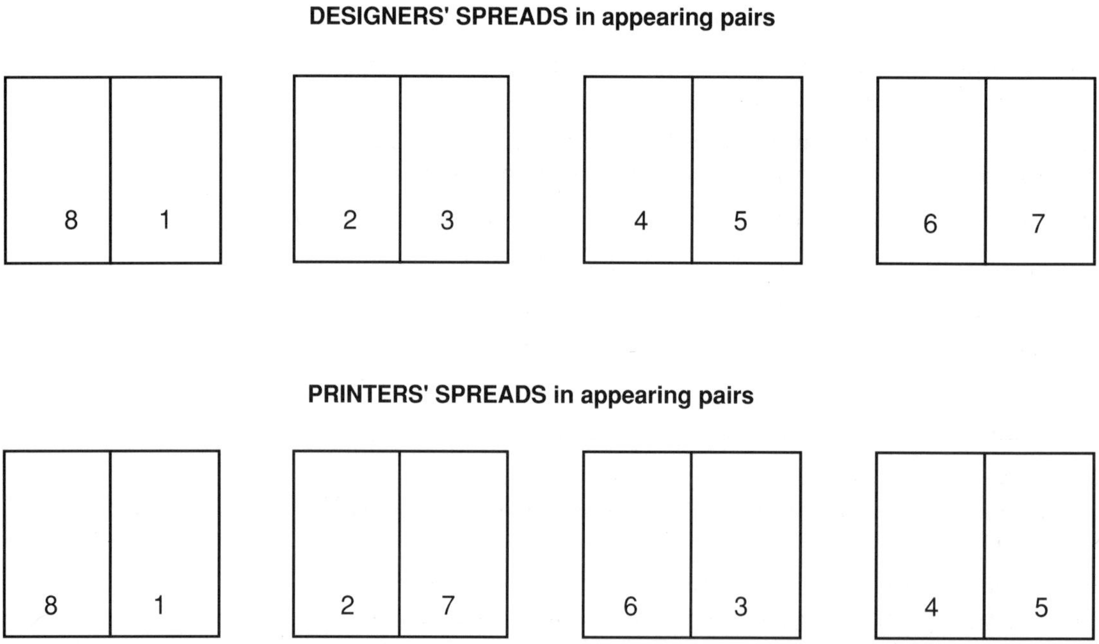

DESIGNERS' SPREADS in appearing pairs

8	1		2	3		4	5		6	7

PRINTERS' SPREADS in appearing pairs

8	1		2	7		6	3		4	5

Figure 9.1: Example of designers' and printers' spreads in appearing pairs

Print specifications and standardisation

In the past it has been considered reasonable that there would be a different printed result when work was printed on different presses, but would otherwise, under identical conditions on the same substrate, produce the desired predictable result. With the growth of multi-coloured printing and the developments and advances in printing presses and the processes, there has come an increasing need to make the finished product more predictable by the establishment of agreed parameters of print specifications.

The major variables which can affect the appearance of printed colour reproduction are in the areas of *graphic repro, plates, paper, water and damp* (for offset litho only), *ink*, and *printing processes*.

To achieve consistent, high-quality printed results to the satisfaction of customer and printer alike, it is necessary to identify and control these variables within agreed tolerances. The resultant findings are then expressed in the form of *print specifications*.

A standard reproduction print specification suitable for a magazine to be printed by heat-set web-offset is as follows:

1) *Film required* - positive film to be 0.004" wrong-reading emulsion-side up. All to be hard dot film with no strip-ins or patch-ups.

2) *Screen ruling* - 47 lines per cm.

3) *Screen angles* - black 15°, magenta 45°, cyan 75°, yellow 90°.

4) *Film identification* - all films to be clearly identified for colour.

5) *Halftone range* - on coated stock as highlight 5%, shadow 90% maximum.

6) *Proofing sequence* - cyan, magenta, yellow, black.

7) *Ink density* - yellow 0.90, magenta 1.3, cyan 1.3, black 1.8.
Tolerance +/- 0.10. All readings measured on a Gretag D186 densitometer.

8) *Printing control strips* - all proofs and progressive must carry colour control targets, preferably Gretag CMS3, positioned across the line of inking. These must be included on all proofs. Only original strips supplied by the manufacturer must be used.

9) *Dot gain* - this must be allowed for during the reproduction and proofing stages. The recommended densitometer readings are as follows: on Gretag CMS3 - 40% tone to record 18%; 80% tone to record 12%; tolerance +/- 2%.

10) *Undercolour removal* - combined colours total of 260 to 300%, lower to upper limits.

A more comprehensive list and explanation of print specifications is included in the IFPP *(International Federation of Periodical Press)* publications *Specifications for European Offset Printing of Periodicals* or *Guidelines for the Reproduction of Halftone Separations* and *Pre-proofs for Gravure Magazines in Europe.* Of particular significance in establishing image quality specifications for newspaper colour reproduction is the ROP *(Run-of-Paper)* publication *Colour Quality Guide* published jointly by the Institute of Practitioners in Advertising and the Creative Services Association.

Colour bars or strips

To ensure continuity between colour proofs and the printed sheet it is necessary to use *colour bars* or *colour strips* so that accurate and meaningful quality control comparisons can be made. These are used so that individual colours can be densitometrically measured; in the actual printing process, the process colours are superimposed and, therefore, cannot be measured separately.

Colour strips or bars are available from many different sources including *Gretag, GATF, Hartmann, FOGRA* and *Brunner.*

GATF allows a simple method of detecting changes such as dot gain, dot loss (sharpening), lateral and circumferential slur, and doubling by visual inspection without the use of a densitometer; FOGRA permits visual and instrumental control of colour strips, whereas Gretag, Hartmann and Brunner require the use of a densitometer to establish measured values.

In collaboration with System Brunner of Switzerland, Du Pont has developed a standard control strip based on the unified European offset colour scale which is compatible with FOGRA recommendations regarding standardisation. The Cromalin Eurostandard control strip contains the same elements used for checking the actual printed result. This means that values derived from the proof can be transferred to the actual print run where comparison can then be made.

Measuring the colour control strips of a proof and the printed result with a densitometer introduces the element of *objectivity* which is essential to print standardisation.

The *density* and *dot gain* of printed images are two of the most important areas which must be controlled and taken into account when predicting the final printed result.

Figure 9.2: Gretag CMS3 colour control strip

Figure 9.2 illustrates the Gretag CMS3 colour control strip consisting of *solidtone patches, halftone patches, trapping patches, circular elements, grey balance patch* and *blank patch.*

The colour control strip should appear on the back edge of the sheet (or web) printed running parallel with the direction of inking, so that the measurements taken accurately reflect the inking profile. Each patch or element plays an important role in the quality control and standardisation of colour.

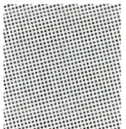

Solidtone patches are used as a measurement of *density in solidtone.* These patches are distributed evenly across the entire length of the strip and measured over the full width of the sheet to gain the most representative readings.

During the printing run, the patches and elements behind key printing areas are measured most frequently with other areas measured less frequently. Colour density values should be in tolerance as outlined, for example, on page 143, item 7, *Ink density.*

Halftone patches are used to determine *dot gain.* The procedure is to first measure the solid-tone patches, then the 40% and 80% halftone patches. Dot gain is expressed as a % gain in relation to the original film strip.

Halftone patches are also used to measure *the relative printing contrast* (from 0-1) which results from the relationship between dot % and density of solid. The higher the printing contrast, the better the gradation and brilliance of printing.

Trapping patches are used to measure *the trapping or acceptance of the second ink layer on top of the first.* The trapping factor is highly dependent on how dry the first colour is when overprinted, and on the viscosity of the ink.

100% represents the complete transference and adherence of the second colour on the first. A 100% transference is only possible when printing *wet-on-dry,* ie on a single-colour press, allowing each colour to dry before the next colour is printed, as against *wet-on-wet* on a multi-colour press.

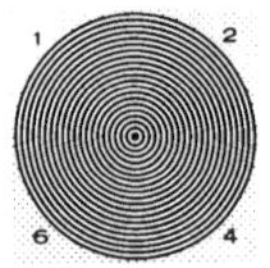

Circular elements are used for *visual recognition of slurring and doubling.*

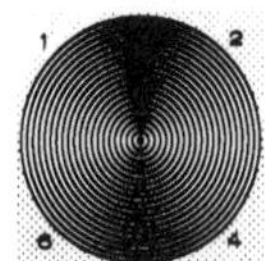

Slurring is identified *where two dark segments are formed in the direction of printing;* the dots in the corner areas on magnification can be seen to be oval in the direction of printing.

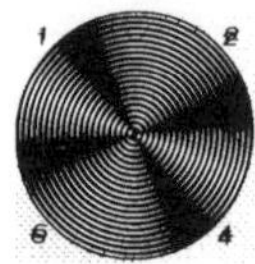

Doubling is identified *where two or more dark segments appear;* these can occur in any direction. The dots in the corner areas on magnification (as well as the numbers) appear doubled. *Dot gain* should never be measured on sheets (or webs) which are slurred or doubled as they will not give a true reading.

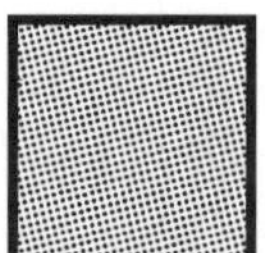

Grey balance patch is used *to monitor the colour balance of the cyan, magenta and yellow.* A neutral grey tone is produced in the grey balance patch if the colour balance of the three basic subtractive colours is correct when fully printed.

The grey balance is used regularly by customers and other individuals without a densitometer to visually check the overall balance of the printing with the adjoining black patch. If the grey balance patch appears cold, ie with a blue cast, it contains too much cyan. If it appears warm, ie with a slight red cast, it contains too much magenta.

Blank patch is used as the *reference value from which to measure* as the densitometer is 'zeroed' on the blank substrate eg paper or board. The densitometer should be set to zero through all filters as the plain substrate serves as the reference value for the density measurements.

Dot gain

Dot gain (dot loss is also possible, but is less common) takes place at almost every stage in the reproduction/printing process.

From the time the dot is first created from a contact screen or digitally, it is liable to change in size in contacting or platemaking. The aim at all stages is to keep the change to a minimum and on the whole this is done, though with more success in the film stages than in the platemaking and printing stages.

Once plates are made and inked, dot gain is much more difficult to control. The dot size will vary with the printed substrate surface and chemistry, damping solution (offset litho only), ink, press design and condition, press room conditions, blanket (offset litho only), and perhaps, most important of all, the press settings and the machine manager's skills.

Dot gain can be considered from two aspects - the *physical* and the *optical*.

Physical dot gain covers any uncontrolled changes that take place between the camera and the platemaking as well as the incontrollable changes on press when the fluid ink is subjected to pressure in the press 'nip'. *Optical dot gain* is caused by the scatter of reflected light which occurs when viewing the printed image.

The proofing material used should match the production substate if further elements of dot gain are to be avoided.

Total dot gain, a combination of physical and optical gain, can be as high as 30% in the midtone (50%) dot areas in web-offset litho printing, so it is obvious that, without a considerable degree of standardisation, no consistency can be achieved.

Every press has its own dot gain characteristics. While other print variables such as ink trapping and solid density can be controlled on the press, dot gain is more difficult to control. Having controlled the dot gain as far as possible for a particular press, the printer should then seek the remedy at the repro stage rather than by further attempting to change the characteristics of the press.

Apart from adjusting the proof press to match the production press as closely as possible, exposure during platemaking should be controlled to produce proofing plates with a dot that will print the same size on the proof press as the final press will achieve from press plates. In addition, proofing inks can be formulated to give a dot gain comparable to that given by the final production inks.

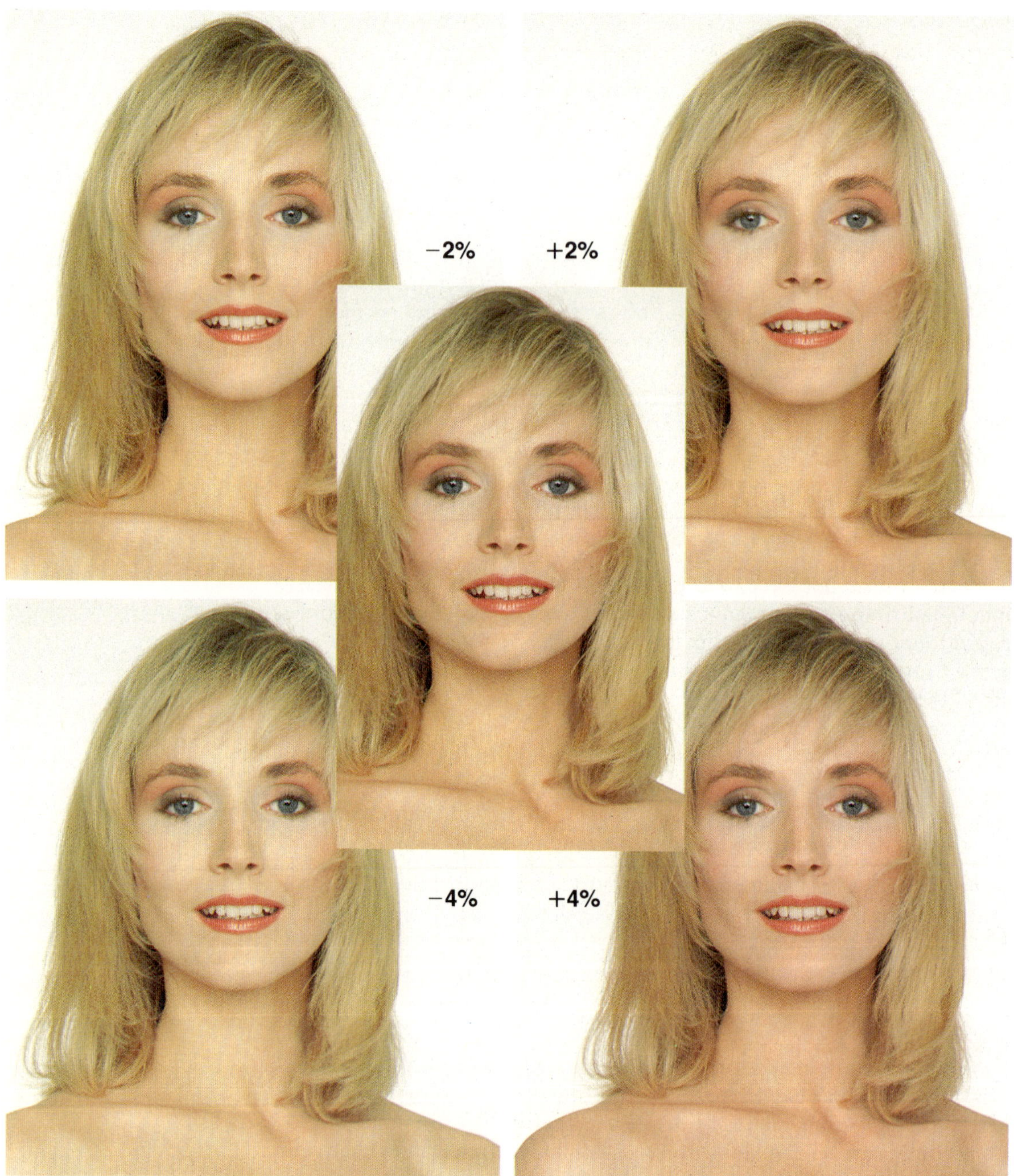

Figure 9.3: Effect of dot loss and dot gain on the printed result

Reproduced by courtesy of Bespoke Publications Ltd and Du Pont (UK) Ltd from chapter five, *Colour Proofing*, of the series '*Colour Concepts*'

Figure 9.3 illustrates how even a small change in dot size changes the overall printed result compared to the desired result.

The centre illustration represents the *reference standard,* ie agreed closest reproduction to the values of the original. All other illustrations are reproduced with varying amendments to the magenta dot. The illustration in the top left hand corner is reproduced with a 2% dot loss; top right hand corner with a 2% dot gain; bottom left hand corner with a 4% dot loss and bottom right hand corner with a 4% dot gain.

The wider band of +/- 4% often represents the tolerance band for standardised offset printing, whereas the +/- 2% tolerance is being adopted by more and more printing companies and print buyers in a drive towards higher quality and more controlled printing standards. Dot gain tolerances are given as +/- 2% in the print specification example for a heat-set web offset magazine on page 143, item 9, *Dot gain.*

Quality control

Quality control in printing, as with any other industry or process, can generally be judged by how close the final product (the printed result) meets with the required set specification (the original used for reproduction). As indicated previously, the printing industry has for some time adopted quality control systems based more and more on standardisation and objective testing methods, rather than subjective assessment which varies from one individual to another.

It must, however, be acknowledged that the human eye is the final arbiter of quality in printing so a visual check between the originals, proofs and printed copies is essential to ensure that all the quality control checks throughout the printing process have resulted in the desired result within the set print specifications.

Quality control aids

The essential aids to quality control employed by the printer include the following:

Register, trim and fold marks

These are reproduced in the excess margins of the printed sheet (or web) so that they are cut off in the finished product.

Control strips, covering:

Graphic reproduction, such as the production of first generation film or contacting to second generation film controlled by guides and gauges used to optimise and control exposures.

The use of a step wedge with known densities, as outlined below, helps to ensure the production of film and other photographic material to the required, consistent standard.

Platemaking, which should include an original strip, such as the *Stouffer Continuous Tone Sensitivity Step Guide*, which is based on a standard exposure time (at the correct light intensity level) correlating with the recommended step on the strip.

Proofing and printing, which need to be compared in a consistent way and this requires the use of a colour control strip such as the *Gretag CMS2* shown as *figure 9.2*. Visual and densitometric readings will allow the colour control strip on the proof and the printed product to be compared.

Densitometers

The use of a densitometer is essential in establishing and monitoring good print quality control standards. A densitometer is a precision instrument that shines light on a selected sample patch and measures the amount of light reflected from it under controlled conditions. The reflected light from the measured patch is broken into the red, green and blue portions of the visible spectrum.

Spectrophotometers

Spectrophotometers, which link colorimetry with densitometry are becoming more popular in printing measurement and comparison, as the move towards the use of special colours in addition to four-colour process gains momemtum. The use of spectrophotometers allow the calculation of measurement values from the colour spectrum. The most common colour measurement parameters are the *CIE colour measuring system, CIELAB colour space* and *CIELUV colour space.*

On a spectrophotometer such as the *Gretag SPM 100,* the measuring parameters can be indicated on the LCD *(light crystal display)* in three different ways - as a numerical value or graphic representation; as a reflection or density spectrum; and finally as a colour locus in the different colour measuring systems as mentioned above. The comparative measurements indicate deviations from the reference value or standard.

Standardised viewing booths

The use of standard viewing booths ensures that originals, proofs and printed products are viewed under standard viewing conditions such as *BS950 Part 2* and/or customer's specific requirements in special cases - see *Chapter 7, Creating printed images,* pages 118 to 120, for an outline on proofing.

Colour management systems

These are systems which have been developed to respond to the need to provide true colour portability and device independence across an increasing array of colour output required from a host colour input source. Examples of colour management systems are *Mac ColorSync, Agfa FotoFlow, Kodak YCC* (used on the Kodak Photo CD systems) and *Kodak Colour Management System.*

In colour reproduction it is essential to control and predict the results from the initial stage of scanning the originals, through the colour monitor used to manipulate and view the work, followed by the intermediate reproduction stages of film, proofs and finally the printed result - colour management systems set out to perform this function.

To assist in the understanding of colour management systems a brief outline of the *Agfa FotoFlow* system is set out below plus, on the following page, *figure 9.4* shows the reproduction of the same colour picture through a thermal transfer printer and offset litho printing machine *with* and *without* the use of FotoFlow:

Agfa FotoFlow

The system consists of a family of four software modules:

FotoTune is the colour calibration member of the system. It creates the *ColorTags* for the input and output parts of the production chain, combining them into a *ColorLink* which translates colours from the input device through a device-independent colour space to an output device with high-quality colour fidelity. By the the use of ColorLinks the colour behaviour of the entire reproduction and printing production chain can be controlled.

FotoReference is the technical colour reference part of the system. It contains standard input targets on transmission and reflective material overcoming the variances inherent in photographic dyes.

FotoLook is the scanner interface 'operating system' where the operator can refer to automated controls or access a variety of features such as highlight/shadow and exposure controls and min/max density controls, enabling a high-quality level of scans. In addition there is access to a colour preview where alterations in black and white points are made in *RGB* and *CMYK.*

FotoScreen is the output member of the system, eliminating the necessity of having images processed by a Postscript RIP before outputting them to an imagesetter. The system utilises *Colorlinks* to transfer colour to the imagesetter and the required output printer by matching the scanner or monitor colour space with the device dependent colour space of the output without loss of quality.

Thermal Transfer Printer

With FotoFlow applied

Without FotoFlow applied

Offset Litho Printing

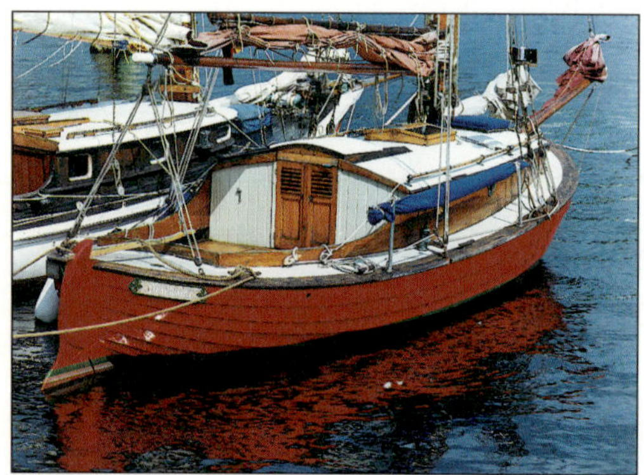

With FotoFlow applied

Without FotoFlow applied

*Figure 9.4: Reproduction of a colour picture through a thermal transfer printer and offset litho machine **with** and **without** the use of the Agfa FotoFlow colour management system*

Reproduced by courtesy of Agfa-Gevaert NV

10 Printing processes

Although new developments and new technology have been applied in recent years at an accelerating rate to how printed matter is produced, we still rely heavily on printing processes whose origins are in most cases well over a century old. Offset lithography, flexography, letterpress, gravure and screen significantly dominate mass market printing; electronic printing, developed over the last decade or so, with its inherent links into digital and multi-media make this an area of obvious future growth. It is likely, however, to remain suited to specialist applications and short-run printing market sectors.

The topics covered in this chapter are: comparisons between the different conventional printing processes; advantages and limitations of the main printing processes; machine printing preparatory and make-ready costs; machine printing running speeds; range of available printing machines; new technical developments in machine printing; suitability of printing processes to different classes of work and the development of electronic printing.

There are five main printing processes clearly identified by the different way the image areas are printed onto the substrates.

Offset lithography is characterised by the *planographic process of printing* where the balance of ink accepting/water repelling image areas and water accepting/ink repelling non-image areas is crucial to the successful operation of the process. Offset litho inks are highly pigmented as they split twice during the offset process, ie from plate to blanket and from blanket to substrate.

Flexography and *letterpress* are both forms of *relief printing* where the image areas are raised above the non-image areas. Ink is retained only by the image areas with the non-image areas remaining uninked. Flexographic inks are liquid inks, water or solvent-based; letterpress inks differ in that they are paste inks of a much thicker viscosity.

Gravure is a form of *intaglio printing*, ie the image areas are recessed below the non-image areas which are flush with the surface of the printing cylinder. Gravure inks are liquid inks and fill the recessed cells before the ink is transferred to the substrate during printing.

Screen is a *stencil printing process* where image areas are clear and unblocked on the screen mesh carrying the printing stencil, whereas the non-image areas are filled or blocked in the required areas. Ink is forced through the clear, unblocked image areas by a squeegee onto the substrate. Screen inks are available in a very wide range of varieties to suit the particular printing requirements.

Comparisons between the different conventional printing processes

Offset lithography is an extremely versatile process, used for a remarkably wide range of printed matter.

Since the impression of the image is first transferred from the plate to the rubber-covered cylinder and then to the substrate to be printed, the process has a further advantage of being able to print satisfactorily 'difficult' and unusual substrates, such as packaging material, bookbinding cloth, or even the silk or satin that is sometimes used in the production of padded greetings cards.

Letterpress is also a versatile process, although nowadays mainly restricted to short-run work, overprinting or specialist reel-fed work.

Type, blocks and plates, can be made up in formes where the different parts are interchangeable and moveable, so allowing alterations and last-minute adjustments or alterations to be made.

Flexography has developed into a specialist printing process, mainly on a web-fed basis for products such as self-adhesive labels and flexible packaging, where increased printing speeds and improving quality has allowed it to claim an increasing share of printing market sectors.

Photogravure is usually considered an ideal process for the reproduction of illustrations in monochrome or in colour, often on cheaper grades of substrate which could not be matched by other printing processes.

The process provides a considerable depth of colour in the shadows and a characteristic softness and delicacy in the middle and light tones.

The cost of preparing gravure cylinders is extremely high, often in excess of £1000 for one cylinder, ensuring the process is generally used for very long runs.

Screen printing, as with flexography, has undergone considerable development and change in recent years, and is now undoubtedly producing very fine quality work across a wide range of applications.

It has particular advantages in its ability to print light colours on dark materials and to print on unusual and uneven surfaces.

Advantages and limitations of the main printing processes

Offset lithography

Advantages

Suitable for a very wide range of work, from short- to long-run

Wide range of substrates able to be printed to a high standard

Fine screen ruling and high definition printed work even on relatively coarse substrates

Wide range of printing plate material available to suit different applications and requirements

High definition reproduction of typematter, line and tone illustrations

Ideally suited to integrating with modern pre-press system links such as DTP and DAR (*Digital Artwork and Reproduction*), also camera-to-plate or computer-to-plate

Widest range of presses available of all the printing processes

Machine speeds generally competitive across a wide range of printing quantities

Convenient to store films or plates for possible reprints.

Limitations

Alterations to machine plates impracticable

Care must be exercised in ensuring the printed quality is maintained by adjusting and controlling the ink and water balance

Being very much a 'chemical' based process it is coming under greater environmental pressures to adopt 'greener' processing practices.

Letterpress

Advantages

Versatility in regard to late corrections, changes of illustrations, dealers' names, etc

Numbering machines can be used and printed perforations can be included in type formes

Economic for short runs and overprinting

Ideal for cutting and creasing, also embossing on adapted presses.

Limitations

Gloss coated paper necessary for fine-screen halftones

Storage of bulky type metal or formes expensive

Comparatively lower machine running speeds than offset lithography.

Flexography

Advantages

Changes to plates can be made relatively easily and cheaply, by just replacing the required parts of the overall image areas

Ideally suited to printing web-fed thin plastic materials with in-line press finishing

Considered an environmentally friendly process as it tends to use few chemicals. Water-based inks are used on most printing substrates, apart from vinyls

Simple, generally easy to use process

Variable cylinder cut-offs allow greater flexibility on the length of images that can be printed.

Limitations

Cannot print screen halftones as fine as offset litho, requires a smooth to coated stock to reproduce good screen detail

Not economic for sheet-fed printing so therefore not suitable for general commercial printed products such as booklets and leaflets.

Photogravure

Advantages

A process giving full colour values in reproduction, with rich tonal effects, particularly in monochrome

Four-colour gravure may equal result of six or more colours in offset lithography

High speeds of great advantage in periodical and magazine work, where very long runs are required

High quality printed results, especially in colour work, on cheap grades of substrates which cannot be matched by other processes

Variable cylinder cut-offs allow much more flexibility on available pagination range for publications compared to heat-set offset litho.

Limitations

Preparatory costs very expensive

Alterations to plates or cylinders impracticable

Type matter and fine-line detail is broken up by the cell structure

Storage of cylinders expensive.

Screen printing

Advantages

Suitable for short runs in multi-colours

Low preparatory costs

Light colours can be printed satisfactorily on dark materials or deep colours

Ideally suited for printing showcards, posters and unusual materials such as heavy gauge metal, plastic, glass etc

Limitations

Halftone subjects are limited to coarse screens

Although automatic presses are now available, the process is still restricted to short-run work. Ink drying also generally slows the process down.

Visual characteristics of the main printing processes

It is undoubtedly a considerable asset to be able to determine with reasonable accuracy the printing process by which an item of printed matter has been produced. There are a number of clues which, if they can be discerned, make identification of the process possible, but not necessarily simple. Indeed, there are some jobs which are very difficult to identify, even to those having considerable experience, and the use of a powerful magnifying glass or 'linen tester' is of considerable value.

There are also some pointers which can help identification of the printing process used to produce a certain item of printed matter, either by the accumulation of evidence or by a process of elimination, but as previously stated it is not easy to establish. The illustrations and comments on the following pages have been prepared to assist in this process.

LITHOGRAPHY

Overall even printed result

Very wide range of substrates including coarse textures, can be satisfactorily printed, even when very fine halftone illustrations are reproduced

Tonal effects obtained by the use of mechanical tints or halftone dot.

LETTERPRESS

Thickening of design under pressure, visible especially around the outer edge of the printed areas

Slight embossed effect usually detectable on reverse of sheet, especially with sheet-fed printing

Tonal effects obtained by the use of mechanical tints or halftone dot

Fine-screen halftones must have substrate with a coated surface.

FLEXOGRAPHY

Thickening of design under pressure, visible especially around the outer edge of the printed areas

Tonal effects obtained by the use of mechanical tints or halftone dot

Fine-screen halftones must have substrate with a coated surface

Printed samples often obtained from processed material in reel-form such as self-adhesive labels, plastic and paper wrappings.

GRAVURE

Wide range of tonal values is possible, giving an effect of continuous tone-like quality (especially in four-colour process work)

Under a magnifying glass the 'screen pattern' is seen to be of a regular square formation (showing uniform cells)

Because of the screen pattern, which appears over the whole of the printed image, fine-line work and text matter appear to be broken when examined with a magnifying glass.

SCREEN

Thickness of ink film is usually more apparent than in other processes, especially where solid colours are printed upon one another

Because of the use of the screen as a support for the stencil, small lettering and fine-line work appear to be broken when examined with a magnifying glass

Halftone subjects are reproduced with a fairly coarse screen.

It is not always possible to be certain which is the most economic printing process to use in any particular application. All printing processes consist of two main contributing factors - *make-ready or preparatory* and *machine printing running speeds* - it is therefore necessary to consider these areas when trying to assess which printing process should be considered.

Machine printing preparatory and make-ready costs

In all printing processes certain costs must be incurred in the preparation stages - that is, all the work involved in preparing the plates, blocks, cylinders and stencils up to the stage at which printing the required number of copies can commence. These preparatory costs will be the same whichever quantity may be required, whether the run is to be 500 or one million copies.

Offset lithography, with its extremely wide range of available press types, both sheet- and web-fed, varies considerably in the cost of plates used and the time required to set up a machine for printing, but nevertheless, overall, it is considered a relatively quick and inexpensive process in terms of preparatory and make-ready costs.

Offset plate material is available in paper, plastic and metal, resulting in plates which can be made in a matter of minutes, and in the case of paper plates, can cost under £1 per A3 size plate from camera-ready copy.

The printed work suited to paper plates is limited short-run, line and coarse screen.

Multi-metal or baked pre-sensitised aluminium plates, although much more expensive to produce, are capable of reproduction of a million impressions and fine screen work up to, and in excess of, 400 lines per inch.

Letterpress has lost ground to the other printing processes as hot metal setting became uncompetitive compared to photocomposition and imagesetting, which are much more suited to the requirements of offset litho, flexography, gravure and screen.

The preparation of letterpress type and/or plates into pages, plus subsequent imposition into machine chases or formes, as well as makeready times, is much more time-consuming and costly than offset litho.

Flexography, along with rotary letterpress, has benefited from improvements in photopolymer plates, and generally the costs of plates and machine make-ready are reasonable, although normally not as low as offset litho.

Gravure has the most expensive set-up costs of all the printing processes, with very expensive cylinders and relatively time-consuming makeready procedures.

Screen has generally low preparatory costs with simple screen stencils being produced inexpensively and quickly.

Make-ready of presses is also quick and relatively inexpensive.

Machine printing running speeds

The output of printing machines varies quite considerably, not only according to the type and age of the machine, but also according to the kind of work being produced, the substrate used and quality standard required - and, of course, the skill of the operator. As a general guide the following average net output speeds are representative for the different printing processes:

Offset lithography

Sheet-fed - 3000 to 10 000 copies per hour
Narrow-width web - 300 to 650 feet per minute
Heat-set and cold-set web - 15 000 to 30 000 'cut-offs'/impressions per hour.

Letterpress

Sheet-fed - 1500 to 2500 copies per hour
Narrow-width web, with in-line finishing - 60 to 250 feet per minute
Large-width web - 350 to 750 feet per minute
Rotary large format - 10 000 to 40 000 copies per hour.

Flexography

Narrow-width web, with in-line finishing - 60 to 250 feet per minute
Large-width web - 350 to 750 feet per minute
Rotary large format - 650 to 2000 feet per minute

Gravure

Sheet-fed - 3000 to 8000 copies per hour
Large-web format - 35 000 to 50 000 copies per hour.

Screen

Sheet-fed - 100 to 4000 copies per hour
Reel-fed - 350 to 750 feet per minute

If a large number of copies are required, say in the region of 750 000, the higher preparatory costs involved in producing photogravure cylinders, for example, become less significant, since the total cost of production - preparatory costs plus machine running costs - will be lower. At the other end of the scale, since small offset and screen printing preparatory costs are relatively low, these processes are usually most economic for small quantities.

Range of available printing machines

Offset lithographic presses are available in a wide range of both sheet- and web-fed formats.

Sheet-fed presses vary from the small offset range of sizes 375 x 500mm and less, mainly in one or two colour/perfector presses which are often fitted with additional facilities such as numbering, perforating and scoring. Larger sheet-fed presses are available from SRA2 (450 x 640mm) to in excess of 1200 x 1600mm in single colour, two- and four-colour, up to eight colours.

Web-fed offset litho presses are available in three main types:

Narrow width presses with fixed or variable cylinder circumferences varying from 17" to 28" approximately, in from one- to four-colour and above configurations.

Heat-set, which is a process where the ink is dried by exposing the printed web to intense heat, then cooled and dried over chill units, has a wide range of presses varying in size from narrow mini-web format - ie eight-page (A4) up to wide-web 64-page (A4) in mainly four and five colours, with specialist presses available in excess of 10 colours.

Cold-set web offset presses, where no accelerated drying of the ink takes place, range in sizes up to and in excess of 1680mm web width, cut-off 630mm and 12 reel infeeds.

Letterpress machines, although representing a declining print market area, are still to be found in a wide range of models.

Many printing companies have retained flat-bed platens and cylinders for a limited amount of printing, but mainly for non-printing activities such as cutting and creasing, embossing, numbering and perforating.

Rotary letterpress machines are now more popular than sheet-fed presses and are available in wide-web, large cylinder and narrow-web small cylinder configurations up to and in excess of 10 printing units.

Narrow-web small cylinder presses are popular for printing self-adhesive labels.

Flexographic presses are mainly web-fed and cover a very wide range from narrow-web widths of 70 to 230mm up to very large width units of up to 2.5 metres, often with six colours or more plus unwind/rewind mechanism - repeat printed images varying up to 1250mm.

Gravure presses are available as sheet-fed format in single, two, four and more colours, often used for specialist work such as printing on metallic foils, or with fluorescent and metallic inks.

There are a range of very large web-fed presses of web widths up to 2400mm, 16 print units and multiple reel stands.

Screen printing machines are predominantly sheet-fed, available as hand-fed, semi-automatic and automatic in sizes up to and in excess of 1100 x 1500mm.

Rotary machines are also available along with specialist and carousel types.

New technical developments in machine printing

In recent years there have been tremendous advances and improvements in reducing set-up times on the printing presses and increasing the running speeds, as well as maintaining, if not improving upon the print quality.

Chapter 7, Creating printed images outlined the advances in pre-press, up to surface preparation: all printing processes have benefited from this application of new and improving techniques.

Just as the pre-press area has embraced the computer to help introduce productivity levels unattainable by previous methods and practices, so printing processes and presses are reaping the rewards of applied computer controls and applications.

Press automation is improving apace in the field of sheet- and web-fed printing.

On sheet-fed offset litho presses pre-set facilities are now available which automatically set up the machine for a new job by resetting sheet size and impression pressure plus automatic blanket wash and ink unit cleaning. Automatic and semi-automatic plate changing is also an option.

Multi-colour presses, both in sheet- and web-format in all the processes, when available, are relying more and more on computerised press control of ink and register, as well as remote service diagnosis improving print consistency and efficiency.

Computerised press management systems are being developed to link into pre-press and print finishing, also into an organisation's MIS *(Management Information System) - see Chapter 3.*

Additional facilities are also being added to presses to improve the value of the printed product, such as in-line coating, accelerated drying and a wide range of in-line finishing.

Suitability of printing processes to different classes of work

Each of the printing processes has particular properties, characteristics and associated costs which make it more suitable for certain classes of work than others.

It has to be acknowledged, however, that there is a considerable amount of common ground where two or more printing processes may regularly be used to produce a certain printed product, eg books printed by offset litho, flexography and letterpress, newspapers by offset litho (cold-set), flexography and letterpress, periodicals printed by sheet-fed, heat-and cold-set web offset, also web-fed gravure.

The comments made below are given as a general guideline rather than a definitive statement on the suitability of different printing processes to different classes of work.

Offset litho

In terms of market share this is the largest and widest ranging of the printing processes.

Small offset
Generally short-run work, up to 10 000 copies of stationery-type products such as letterheadings, business cards, overprinting of envelopes, pads, sets; leaflets and booklets etc.

Larger-size sheet-fed
Generally most competitive in print runs of up to 50 000 copies, although in certain circumstances it can also prove economical in print runs up to 250 000. Range of work includes books, booklets, brochures, cartons, catalogues, folders, magazines, annual reports, instruction manuals, posters and leaflets etc.

Narrow-width web
Mainly specialist work such as business forms, and to a lesser extent labels, plus if sheeter and dryer is fitted, general commercial work normally in multi-colours.

Heat-set web
Generally most competitive in print runs of above 50 000, but reductions in set-up times and material wastage, especially on mini-web presses, can make run lengths as low as 10 000 competitive. Range of most suitable work covers magazines, holiday brochures, catalogues, brochures and direct marketing products. Paper stock range producing folded printed sections or products is normally restricted to between 40 to 135g/m^2, higher if folding is not required on-line.

The main competitor to heat-set web offset in long-run colour work is web-fed gravure, although in up to 250 000 copies heat-set web offset tends to hold a cost advantage.

Cold-set web
Mainly suited to newspaper and newspaper-type products; longer-run paperback books and directories in spot colour and four-colour process.

Letterpress

Sheet-fed
Restricted range of short-run work such as business cards, letterheadings, leaflets, booklets and posters in mainly one or two colours. Extremely popular for 'non-printing' operations such as cutting and creasing, die cutting, embossing, numbering and perforating.

Narrow- and *larger-width web*
Specialist work such as books and self-adhesive labels, flexible and rigid packaging, limited range of continuous stationery such as computer listing paper.

Rotary large format
Similar range of printed products to flexography. Main range of work includes regional newspapers and newspaper-type products, also longer-run paperback books and directories in black only or black and spot colour.

Flexography

This is predominantly a web-fed process, suited mainly to specialist or niche printed markets such as reel-fed labels, newspapers, flexible packaging such as food wrappings, carrier bags and rigid packaging such as cartons and collapsible corrugated cases.

Gravure

Sheet-fed
Suited to specialist work such as printing on metallised and other substrates to produce high quality decorative effects in gold, silver and flourescent colours.

Web-fed
This main application covers a wide range of general commercial products.

Gravure is especially suited to work in four-colour process on relatively cheap, smooth mechanical papers in quantities of 250 000 or more, such as magazines, mail order and catalogues.

In addition there are a wide range of specialist products such as security printing including stamps and cheques; board packaging products such as folding box cartons for food and cigarette industries, also printed video cases; flexible packaging such as printed cellophane and polythene in food wrapping, display and protection.

Screen

Sheet-fed
As the process is best known for its ability to print a thicker ink film than any other printing process this makes it ideal for printing light coloured inks on dark coloured materials, also onto awkward, rough surfaces, uneven and moulded shape surfaces. Examples include posters, showcards, printed circuits, T-shirts, printing on cloth, vinyl, metal, glass and plastic.

Rotary/web-fed
Specialist area of the process used for self-adhesive labels, scratch-off lottery tickets, packaging, transfer printing, security printing, direct mail and high quality greetings cards with die-cutting and additional finishing requirements.

Development of electronic printing

Although all the main conventional printing processes, covered previously in this chapter, are relatively undisputed in terms of volume and value produced each year, electronic printing in its many forms is now also established as a major player in printing and print-related products.

Electronic printing covers the processes used to create images on paper or other substrates, generated from digital data, without the use of a physical image carrier or plate; apart from some copiers which may work from a paper master, at least in the initial stages.

It is generally accepted that electronic printing is ideally suited at present to *on-demand printing* with extremely short-lead times. This is due to the ease with which electronic printing systems can link the areas of pre-press, printing and often finishing into one integrated process.

The images to be printed are generated in a digital form, often on an AppleMac, PC or other host computer system; they are then downloaded onto the host printing unit via a link which can either be part of an on-line networked system or the data transferred off-line by a storage media such as a disk or tape. The print unit is then set up for the required copies and if a finishing unit is included the job is collated and bound.

Whereas conventional printing processes are mostly suited to producing the same printed image time after time in volume, electronic printing has the advantage and facility to produce a different image each time, and so can produce an original copy each time if required. This is due to each of the conventional printing processes requiring a relatively expensive plate, cylinder or stencil for each separate image, whereas electronic printing can literally produce a new image at the flick of a switch.

Range of available electronic printing systems

Electronic printing systems are now available which can produce black only, spot-colour or full-colour, either for short-run work or specialist applications.

Photocopiers/printers

The new generation of copiers, both black-and-white and colour, is the development which threatens certain conventional printing market sectors presently held predominantly by small-offset. Copiers now use technology which offers much higher resolution than previously, with scanning-in resulting in excellent quality, even with photographs.

The more advanced system copiers are actually changed into a printer, rather than a copier by connecting up to digital data from an AppleMac, PC disk or are linked/networked to a host computer.

The *Xerox Docutech Production Publisher* is a very highly-specified copier/printer system producing 135 A4 pages per minute at 600 dpi, and when connected to a network server it links up with most established hardware and software, having over 1Gb of internal storage for job set up, job contents and retention of a substantial library of logos, etc.

Most photocopiers, whether linked to a system or not, can produce spot colour work by simply using coloured toner powder.

Colour copiers have now advanced to the stage where they can be linked up to AppleMacs and PCs and used as an input scanner as well as an output copier.

Digital colour printing

New systems are being developed to bridge the gap between colour electrostatic copiers/printers and conventional printing in terms of the quality achieved by conventional printing; also the facility to print on a wide range of substrates which is not available on most copiers, but also to gain the versatility of unique digital imaging.

Systems that fall into this category are the *Heidelberg GTO-DI (Direct Imaging) press* which accepts digital data to produce plates on press which are 'waterless', but nevertheless use standard printing inks. A wide range of substrates can be used and the printing speed is up to 8000 sheets per hour, with maximum printed size of B3 (360 x 520mm).

Further systems are the *Indigo E-Print 1000*, which again accesses digital data to create the printed result from an imaging and blanket cylinder using a special ink called Electroink. The maximum printed area is A3 (297 x 420mm), at 4000 A3 size images per hour, with up to six colours perfected, collated and bound into a saddle stitched booklet if required. The *Chromapress* is a further digital colour printing system, but this time prints from a continuous web, sheeting off at a maximum size of A3.

Specialist applications

This range of electronic printing systems includes ink-jet, which is mainly used for personalisation and specialist printing on packaging and direct marketing products, labels and cans. The major uses of electronic printing, including ink-jet, laser, electrostatic, thermal, bubble-jet, thermal wax, and dye sublimation is for very short-run work, possibly only a few copies up to a few thousand copies, but also in the area of providing proofs from a DTP/digital pre-press systems.

11 Print finishing

Print finishing is the operation or series of operations carried out to ensure each item of printed matter is processed into the final form suited to each customer's specific requirements. It is an area that has remained relatively unchanged in terms of production techniques in recent times compared to pre-press and printing operations, which have seen dramatic changes through the impact of computer applications.

Due to the wide variety of work undertaken by printers, versatility and flexibility of finishing operations undertaken are essential, with manual bench work co-existing alongside increasingly sophisticated equipment.

This chapter covers the following areas: dummies; range of print finishing operations - in-line, on-line and off-line finishing; cutting and trimming, folding, gathering, collating, insetting and inserting, wire stitching, thread sewing, perfect or adhesive binding, cutting and creasing, die cutting and label punching, embossing and foil blocking; bookbinding - softback and hard cased books, stationery binding, plastic comb and spiral wire binding; print finishing operations associated with different types of printed work - general/ jobbing or specialist; coatings and finishes; laminating and graining; binding and finishing techniques.

Most printed items need finishing in some form or other to turn them from a sheet or reel of the selected printed substrate into, for example, a bound brochure, booklet, continuous stationery set of invoices, direct mail shot, label or a carton.

Print finishing needs to be carefully planned, due to the wide variety of processes which can be included, either applied separately, or as part of a complex in-line process.

Dummies

It is often advisable to prepare 'mock-ups' or 'dummies' to illustrate and demonstrate the proposed finished result before being committed to expensive materials and production processes which may not result in the desired objectives.

The procedure involves making-up a dummy of the job with the correct materials and specification to ensure the finished product will achieve the required results, eg qualify for the lowest possible postal weight/ charges; also to check that the paper and folding sequence chosen provides a cost-effective, attractive and purposeful result.

In addition, it allows the opportunity to check if a magazine will need to be perfect bound rather than saddle-stitched, owing to the high pagination making it particularly bulky. Similarly, it will help to see if a carton design or presentation is practical as well as attractive in concept.

Dummies will allow all parties in the print chain - the customer, print designer and printing company - to ensure costings, quality considerations and 'technical practicalities' can be checked out thoroughly before committing to the major production run. Dummies will often be prepared up to and including the proofing stage, in the form of 'test' or 'advanced' copies, to allow a final check and highlight any previously unforeseen problems.

The vast range of the different operations available often means that printing companies call upon the services of print finishing trade companies for specialist or outwork services. Print buyers also have the opportunity of buying the services of trade print finishers direct - see *Chapter 13 Purchasing procedures.*

Range of print finishing operations

In-line finishing

In recent years there has been a growth in *in-line finishing* where more and more web-fed presses can now deliver the completed job off the press.

In-line finishing is the operation, or series of operations, in which printing machines have the facility for finishing operations to be carried out on the press.

In the move towards automation in print finishing and bindery operations, there has also been considerable growth in automated binding lines where several operations are linked together in an in-line binding configuration - examples of this type are available for saddle stitching, perfect binding and thread-sewn bound books.

The main advantage of in-line finishing is the potential to *print-and-finish* the printed product in one continuous flow-line of operations. The major drawback is that the press running speed is considerably slowed down by the on-press finishing operations, such as the folder.

Set-up times can also be considerable: for example, a complex direct-mail job can take as much as 10 to 12 hours to set up the complete operation before printing begins. However, taking everything into account, it is still often cheaper and faster to finish in-line if possible.

On-line finishing

An alternative method of finishing involves normally free-standing operations such as collators, numbering units, perforating units, folding and wire stitching machines which can be temporarily linked together in an on-line configuration. This is most common in the area of small offset printing units.

Off-line finishing

This is where print finishing takes place after, and separate from, the printing operation. In this area each finishing operation stands alone, with each unit operating to its individual maximum capability without any restriction to its speed from other linked operations.

Few sheet-fed printed jobs are complete when they come off the printing machine, most having to pass through another process or department for completion and therefore off-line is the main type of print finishing used in sheet-fed printing.

An outline is now given of the range of print finishing operations commonly carried out by printers and print finishers.

Cutting and trimming

Single-knife guillotines

Practically every sheet-fed printer uses a guillotine for cutting stock down to the required working size for printing and also trimming to final size.

Single-knife guillotines consist of a back gauge, flat-bed, clamp and guillotine knife with models varying from basic, mainly manually operated machines to semi-automatic and fully-programmatic machines.

Programmatic guillotines allow the operation of a pre-set programme where the guillotine automatically goes through a series of operations where the substrate is cut to the required size in one dimension: when completed, the guillotine operator then turns the sheet around 90°, activates another selected programme and the guillotine cuts to the required size.

Some guillotines form part of an *integrated material handling and processing work-flow*, where automatic airing and fanning devices, jogging units, air-assisted transport system, programmatic guillotine, automatic waste removal, handling and packing facilities streamline the process to a high degree - *Figure 11.1* overleaf illustrates the *Polar system 7* which is an example of this type of equipment.

Figure 11.1: Guillotine as part of an on-line material handling and work-flow unit

This type of equipment would, however, only be used by a printing company with an intensive and regular requirement for high volume throughput of similar printed product lines such as sheet labels, postcards and greetings cards.

Three- and five-knife trimmers

This piece of equipment is used to trim bound work to finished size, ie head, tail and foredge in one operation, whereas a single-knife guillotine would require three separate trimming operations.

It is used mainly by printers who have a requirement to produce a high volume of bound publications such as booklets, magazines, catalogues and books, etc.

The machine can take the form of a stand-alone operation or as part of an in-line finishing operation linked to an automated binding line. When set up as a three-knife trimmer the machine produces one batch of bound publications at a time, whereas as a five-knife trimmer it operates as a two-up finishing unit.

Folding

One of the most common, yet most important of finishing operations, is folding sheets of printed paper or board down to the required format and size.

This can, of course, be done by hand, especially for very short-run work, where awkward shapes or complex folders with gussets and pockets have to be made up from a special die cut shape.

Generally, however, machine folding is used by most printers and print finishers from a wide range of models available to suit a range of applications - from small desk-top size folders capable of only one or two simple folds to large format machines capable of multiple and complex fold configurations. Specialist folding machines are also built into an in-line finishing operation such as the folder-gluer machines used by folding box carton manufacturers.

Folding machines can also be fitted with a wide range of ancillary equipment so that they become a complete in-line finishing unit performing in one pass operations such as pre-scoring, gluing, perforating, folding and slitting to final finished size; and on-line packing units such as strapper and shrink wrapper. Many large folding machines have been adapted or purchased with many of these facilities for printers to undertake finishing of direct marketing products - see page 183.

Gathering, collating and insetting

These are finishing operations necessary when a bound job consists of multiple sections which need to be brought together into complete sets ready for binding.

Gathering is the process of placing sheets or sections in the correct sequence to make up a bound publication such as a book, booklet, magazine or catalogue, etc, or pads of forms consisting of printed leaves in multiple sets.

Collating is checking through printed sheets to ensure they are complete and in the correct sequence for binding. The term collating is often used nowadays to describe the process of collecting sheets into a pre-determined order which, technically, is gathering.

Insetting is the process of placing one section inside another for subsequent binding; it is necessary when there is more than one folded section in, say, a booklet or magazine which is to be saddle-stitched.

Inserting is a completely different operation to insetting, although the two terms can easily get mixed up or confused with each other.

Inserting is the process of placing a loose insert such as a leaflet or direct marketing product between the leaves of, for example, a booklet or magazine without binding it to the main printed product.

All four previously mentioned operations can be carried out as entirely manual operations or automated procedures, also as stand-alone off-line, or as part of an integrated on-line finishing unit.

Wire stitching

Saddle stitching is the cheapest and fastest method for binding (apart from in-line gluing) single or multiple insetted section work for booklets, magazines and catalogues.

The wire or staple is driven through the back fold of the insetted sections and clenched across the centre fold. Saddle stitching permits the bound leaves to open flat, but the thickness should not exceed 7mm, otherwise the finished copy will not lie flat when opened, also the wire stitch may not securely bind the centre folded leaves.

Side-wire stitching is a method of binding gathered work, consisting of single leaves or folded sections, by the wire stitches being driven through the binding edge from front to back.

It has the advantage of stitching thicker jobs than saddle stitching, up to 50 mm, but it has the major disadvantage that when opened the leaves will not lie flat and a plain binding margin of 25 mm to 50 mm is required.

Due to the inherent properties of this type of binding it is mostly suited to bound pads of multiple sets, where the file copy to be retained is held fast in the pad and the perforated copies are withdrawn from the set.

Thread sewing

Thread sewing is a method of binding gathered folded sections where the sections are opened at the centre and thread sewn through each and every section, so that they are bound together by thread and subsequently by glue.

This method results in the highest quality binding in either softback or hardback books.

Perfect or adhesive binding

Perfect or adhesive binding, also known as *unsewn* is the preferred method for binding thicker magazines, catalogues, paperback books and other work for which saddle stitching cannot cope, also where thread sewing is too slow and costly or where side-wire stitching results in a job that will not lie flat when opened. Sections are gathered and collated, then clamped securely while the spine folded area is sawn and roughened off so that the adhesive will penetrate the individual leaves, resulting in a very strong binding.

The previous methods of binding covered, such as saddle and side-wire stitching, thread sewing and perfect binding are again carried out in the industry as separate hand-fed operations or linked to automated binding lines of an on-line configuration.

Cutting and creasing

This is a finishing operation carried out on specialist machines such as *Bobst* used by carton companies, or adapted letterpress printing machines such as *Heidelberg cylinders* by general commercial printers, both of whom have a requirement to produce printed products made up from an irregular shape. Special cutting and creasing dies or formes are prepared by hand or laser to the shape and configuration required.

Figure 11.2: Bobst automatic platen press used for die-cutting and embossing folding cartons

Die cutting and label punching

This is a method of finishing where cut-outs and irregularly-shaped items, such as labels and showcards, are cut-to-shape on punching machines of various sizes by steel label punches of the required pattern. The cutter is driven into the material with the cutting edge on the inside or outside of the shape according to which portion of the sheet is to be discarded.

Steel label punches used with special ram punching machines are more expensive than cutting formes made from cutting rule, but they last longer and can be sharpened further if required.

Die cutting of reel-fed self-adhesive labels is mostly carried out as an in-line printing and finishing process which includes printing, die cutting, waste stripping, re-reeling and slitting of reels to smaller size where required - see page 183.

Embossing

Embossing is created by the pressure of the substrate between a hollow female die, normally made of brass or rigid photopolymer, and a male counterpart. It is produced on a flatbed letterpress or specialist machine where the male counterpart is built up on the bed of the machine.

Layers of paper and thin card are interleaved with paste and built up to form a soft mould which is then pressed into the hollow female die to the correct depth to form the male counterpart. This is then allowed to dry, after which sheets of the substrate are fed into the machine to receive the required embossed or raised effect. An alternative material used to make the embossing dies is plastic, formed by heat-assisted moulding.

Embossing is used extensively in business cards, letterheadings, cartons, high-quality labels and brochure covers, where the raised embossed effect gives a pleasing and attractive finish.

Figure 11.2 illustrates a specialist Bobst machine, as used by carton printers, which can be fitted with an embossing unit.

Foil blocking

Foil blocking has become very popular because of its attractive and brilliant metallic finishes in gold and silver, although it is available in a wide variety of colours.

The foil consists of several laminates bonded to a base carrier film so that when heat and pressure are applied using a relief image, normally of brass, hard plastic or photopolymer, the image area of the foil transfers to the substrate.

Special blocking presses are used in producing specialist printing and print-related products such as labels, cartons and hard-case bookbinding.

Alternatively, hot-foil attachments are available which convert old flat-bed letterpress platen and cylinder presses into hot-foil blocking presses, basically by removing the inking mechanism and replacing it with a heating and foil-feeding unit.

Bookbinding

Bookbinding is a print finishing term which covers a wide range of bound publications including softback books, hard cased books such as edition case and library style binding, plus stationery examples of account books and loose-leaf. Additional forms of binding also include plastic comb and spiral wire.

Softback books

Softback, or *paperback* books as they are also known, are the most prolific method of bookbinding as they provide an economical and practical means of producing short-run and long-run bound forms of a wide range of reading matter.

Perfect binding is used in softbacks where the cost factor is paramount, as this represents the least expensive form of binding a thick, flat-back publication. Thread sewing is the form of binding used for softbacks where durability is important, including reference publications such as text books and manuals which need to withstand regular handling - this book is an example of a thread sewn softback.

Hard cased books

Edition case binding

Today the majority of hardback books are case bound, a style also known as *publishers' edition binding*. This method of binding gets its name from the cover, or case, being prepared separately from the thread sewn sections, the two parts being brought together only at the end of the binding process.

Library style binding

This type of binding differs from edition case in that the cover boards are normally attached to the sewn sections in one integrated operation, whereas with edition case binding it is two distinct operations. It remains largely a hand, craft-based operation, resulting in an extremely strong and durable method of binding which is used for the more expensive books, intended primarily for library use.

Stationery binding

Stationery binding is the name given to the wide range of bound work which is primarily intended to be written in, such as note books, duplicate books, receipt books and cheque books, etc; they tend to be side-wire stitched or saddle-stitched. The style of binding for books and pads which are to be used for written records, of either a temporary or permanent nature, needs to be different from that of books produced for the purpose of reading and reference material. Their size, shape and durability depend on the purpose for which the stationery books are required.

Account books, although within the category of stationery binding, is an entirely separate section of the bookbinding trade.

The construction of an account book is different from that of edition case binding. Account books are rounded, but not backed, having what is known as a *spring back*, which allows the book to open flat for writing when required. The bound leaves are pattern ruled or printed in light blue and red as required; in addition the leaves are normally numbered as well as a leaf index being included.

Loose-leaf binders have tended to supersede sewn account books, which are much more expensive to produce and are more cumbersome in operation.

There are three main types of loose-leaf binders:

Ring binders. Snap rings on a metal tube holds sheets which are punched with corresponding holes. Standard sizes are available with two or four rings at 80mm centres, or with multi-rings.

Post binders. Metal posts, fixed to the cover at their bases and adjustable at the top, hold the sheets which are punched with corresponding holes or slotted to allow removal of sheets without disturbing the mechanism.

Thong binders. Flat thongs, which may be slackened or tightened by a mechanism in the cover, hold the sheets which are slotted to fit the thongs.

Plastic comb and spiral wire are two popular and relatively simple methods used for binding reports, documents and notebooks, etc.

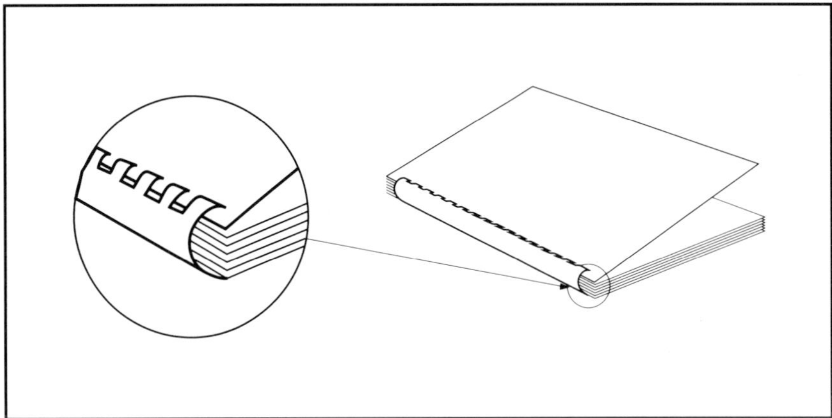

Figure 11.3: Plastic comb binding

Plastic comb binding

Work to be plastic comb bound has to be completely made-up prior to binding.

Gathered sections, single loose leaves or a combination of both can make up the job to be bound, with the made-up job trimmed prior to binding.

After trimming, the book is placed on a punching machine where a multihead punch makes a series of slots on the binding edge, some 6mm from the spine, depending on the thickness of the book and the type of plastic comb being used. The comb-binding machine is loaded with the correct length of pre-formed plastic comb. The punched book is held securely in the machine spine uppermost, the comb is opened and the teeth located in the punched slots; closing the comb secures the sheets.

Work bound in this way opens flat and the spines can be blocked or printed. As with side-stitched and adhesive-bound work, allowance must be made for the loss of the margin on the binding edge due to the method of binding.

Plastic comb bound books may be made up from a variety of materials - for example, sample books of printed plastic, material, colour or paint swatches mounted onto paper or board. Equally, the covers can be simply a sheet of clear acetate, a printed board or a plastic-welded composite.

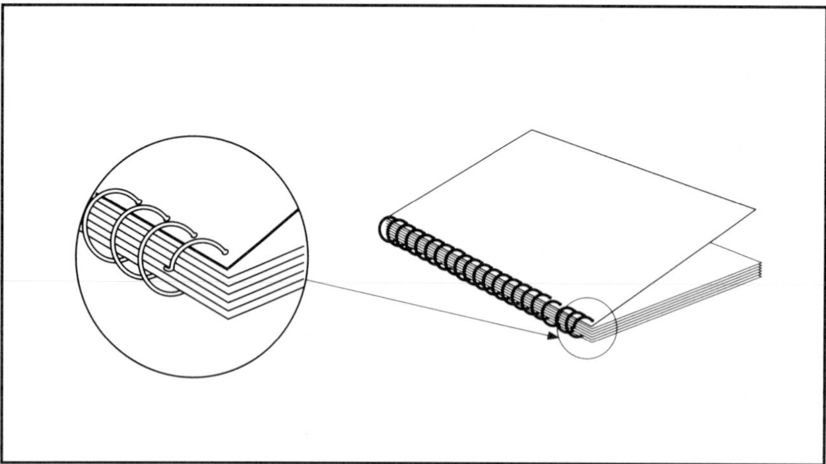

Figure 11.4: Standard spiral wire binding

Spiral wire binding

Work to be spiral wire bound is made up and trimmed in the same way as for plastic comb binding. After trimming, the book has small holes punched in it by a multiheaded punch on a punching machine to accommodate the wire. The binding is then completed on a special machine.

A continuous reel of wire is fed into a former to spiral the wire and space the loops the same distance apart as the punched holes; this is known as the *pitch*. As the leading end of the spiralled wire emerges from the former it passes through the first punched hole at the head of the book and, as more wire is fed, the leading end threads its way through the rest of the holes until the binding is complete, when the wire is cut and the ends turned in at right angles.

Work bound in this way is often used for calendars and notebooks, due to the ease with which it opens and remains flat. As with plastic comb binding it requires an allowance for loss of margin. It is not possible to maintain precise alignment between the pages of a spread as the sheets have a tendency to ride up the wire, causing misalignment across the double page spread.

'Wire-O' is a proprietary system of spiral binding which allows the alignment of each spread to be maintained by using separate binding units rather than a continuous reel of wire.

Print finishing operations associated with different types of printed work

Chapter 1, pages 5 and 6, introduced the subject of 'Printing companies and their customer base' breaking down the different types of printing concerns serving different market sectors into *general* and *specialist*. The range and type of print finishing equipment used by print-related organisations fall into this same pattern.

General or jobbing work

This forms the largest and most comprehensive group of all printers. The vast majority of finishing operations are off-line, including guillotines, folding machines and wire stitching, plus a wide range of miscellaneous, often bench-linked operations, such as drilling, round cornering, numbering, scoring, perforating and padding.

Gathering, collating, insetting and inserting are mainly manual operations with smaller-sized printers.

However, print finishing equipment used by medium-to-larger sized printers producing sizeable quantities of collated sets, booklets, brochures and magazines is likely to be at least partially automated on an on-line or in-line. Examples of the type of equipment used would include a collating machine linked to a padding unit, automated saddle stitching or perfect binding line consisting of section feeding, binding and a three-knife trimming unit.

Specialist work

Periodical printers usually have a more restricted range of print finishing equipment than general printers, but nevertheless tend to use a more specialist and automated range.

Sheet-fed periodical printers use guillotines, folding machines and in-line automated saddle stitching and perfect binding lines with cover feed and insert feeders, plus mailing and packing lines often linked to postal sorting systems.

Web-fed periodical printers do not normally have the requirement for folding machines as the printing presses produce folded sections. The use of a guillotine is often restricted to cutting covers down to size, otherwise the equipment is as for sheet-fed periodical printers, apart from additional handling equipment due to the bulk produced by the web machines.

Newspapers are produced on dedicated in-line presses which print, fold and finish, resulting in a broadsheet or tabloid newspaper. Depending on the capacity of the press, there may be a requirement for a section inserter if the required pagination cannot be produced in one web pass.

Books undertaken by specialist book printers and finishers are mostly produced on specialist equipment, often adapted to individual company needs.

This can either be in the form of free-standing batch processing lines, or integrated mass production flow lines.

In the area of glued/perfect binding lines for thick publications, softback and book block manufacture, the following equipment is used: endpaper gluing, section gathering, perfect binding and three-knife trimmer.

For thread-sewn, case bound manufacture, the range of print finishing machines used consists of: high-speed thread-sewing, book rounding and backing, bookmark inserting, backlining and headlining, high-speed case-making, automatic foil blocking/embossing, high-speed casing-in and book jacketing.

Cartons and packaging products are printed mainly on sheet-fed (offset litho) presses, with some very long run work such as cigarette cartons on web-fed (gravure) presses.

Sheet-fed presses used for carton printing tend to be five, six or more colours with in-line coating and UV or IR drying facilities to enhance the aesthetic appearance and finish of the carton.

Cartons are multiple-image printed on a large sheet and therefore require cutting out, creasing and waste stripping after printing on a large specialist press. This finishing press can also be capable of embossing, foil blocking and window patching or aperturing.

Figure 11.2 illustrates a Bobst automatic platen which is capable of all these operations if the machine is configured to this wide-ranging requirement. The cartons are then finished on specialist folder/gluer machines which, as the name suggests, fold and glue the finished cartons prior to packing in suitable batches.

Labels can be either sheet-fed or reel-fed printed. If *sheet-fed* and of a straight-edged rectangular shape, the labels are trimmed to size on a guillotine, followed by banding and packaging in batches as required. Irregular shape labels will be die-cut or ram-punched to size, followed again by banding and packaging.

Reel-fed self adhesive labels are produced on an in-line machine where the printing and drying units, are followed by a range of finishing equipment which will normally consist of hot-foil, laminating, die cut and waste stripping, to finally re-reeling and/or slitting and sheeting units.

Continuous business forms produced on multi-web-fed presses offer the facility of printing and finishing in-line, where the printed webs, if the facilities are available, can be numbered, file/sprocket hole punched, perforated, slit, glued, crimped, collated, bar coded and block-out de-carbonised, to deliver the finished product in finished reel-to-pack form.

The following equipment is used for off-line finishing of continuous business forms:

Converter/paper processor. This is a simple machine which processes reel stock - normally into pack form - for further processing, or for producing plain paper stock. Operations will include producing fan-folded or pack products, which can be sprocket hole punched, perforated and also numbered.

Pack-to-pack collator. As the name suggests it is a machine which collates pack sets, normally with up to seven loading tables, therefore producing up to seven-part sets. Facilities can include crimping, tab fastening, gluing, numbering and perforating.

Reel-to-reel collator. This type of machine is used where long runs and multi-set work - either cut or continuous - predominates. Collators can be up to eight stations with sets being glued, crimped, crash or conventional numbered as required.

Direct marketing is another range of products which can be produced on web-fed in-line printing and finishing equipment, or sheet-fed printed and off-line finished.

If sheet-fed, the central finishing unit takes the form of a folding machine with a wide range of additional finishing attachments capable of producing a one-piece mailer.

Web- or sheet-fed, the principal finishing applications are folding, impact glue, remoist glue, perforations and personalisation plus additional applications such as die-cutting, window patching, rub-off and aromatic inks, also additional inserts enclosure.

Coatings and finishes

The appearance of finished printed products can be significantly enhanced by the application of a *coating, varnish* or *laminated film* finish.

Apart from looking more attractive, the printed product is given added protection when a coating, varnish or laminated film is applied overall such as on a brochure cover or a carton. Another advantage is the improved aesthetic appeal where spot varnishing is used to highlight and make certain areas appear more prominent.

Coatings and finishes offer the dual advantage of:

a) *design aid* where the surface finish can be 'overall' or 'spot' as glossy, matt or plain

b) *surface protection* where the printed surface is given increased wear and protective properties.

Coatings can be applied to a wide range of printed products; the choice of coatings is influenced by the cost and overall finish required.

In-line coating on a printing machine, or as a separate machine unit is available in three main types:

Varnishing is essentially a 'colourless ink', ie a clear varnish, which is applied like an ink. It gives good moisture protection, has similar properties to ink, but can 'yellow' with age and is generally slow drying.

Aqueous coating consists of approximately 40% solids, 60% water and is applied through the in-line coating system, which is fitted to certain printing machines regularly requiring this facility, such as for carton and sheet-fed label production. Occasionally, aqueous coating is also applied through the dampening unit. It gives a good hard surface with high gloss when required and is fast drying with no yellowing of the printed result as is often experienced with varnish. There is a risk, however, of sheet stretch or shrinkage with lower grammage papers, due to the high water content.

UV (ultra violet) dried varnish or coating. Drying by radiation gives the best finish in terms of results. The whole coated surface changes into a solid state. The result is an extremely high gloss finish comparable to gloss laminating. Instant drying with very high abrasion resistance, but requiring special inks, make this method more expensive than ordinary varnishing or aqueous coating.

All types of coatings cut down, if not eliminate altogether, the need for anti-set-off powder which can be detrimental to the final result. Coating is also used to improve rub resistance on printed matt coated substrates.

Laminating

This is the application of a clear plastic film by wet or dry adhesive over the entire printed surface, resulting in possibly the ultimate finish in terms of protection. The film covers the entire surface and cannot be applied on a 'spot' or selected area basis. It is often difficult to tell the difference between overall surface UV varnishing and laminating - sometimes the only way to be sure is to try and tear the finished result; if it tears easily, it is UV varnishing. Laminating is often carried out as an outwork process through a print finishing trade company, although some printing companies have installed their own facilities.

Hints on film lamination

1) Lamination

The process places a thin plastic film on one side of the sheet. After one-sided lamination it is possible for the reverse unlaminated side of the sheet to either take up or lose moisture, resulting in curling of the substrate, particularly with paper and lighter weight boards. To help reduce this, it is advisable to moisture-proof the material whilst in storage, and to try and prevent exposure to extreme conditions before, during, or after processing.

2) Laminated area

In the laminating process, as reels of film are used and adhesive is applied to the film, it is necessary for the film to run off one of the edges of the sheet being laminated. The width of sheet must, therefore, always be larger than the area of lamination required, and it is normal to have an *unlaminated margin* on three of the sides of the sheet. Two of these edges will be the *printer's gripper* and *sidelay*:

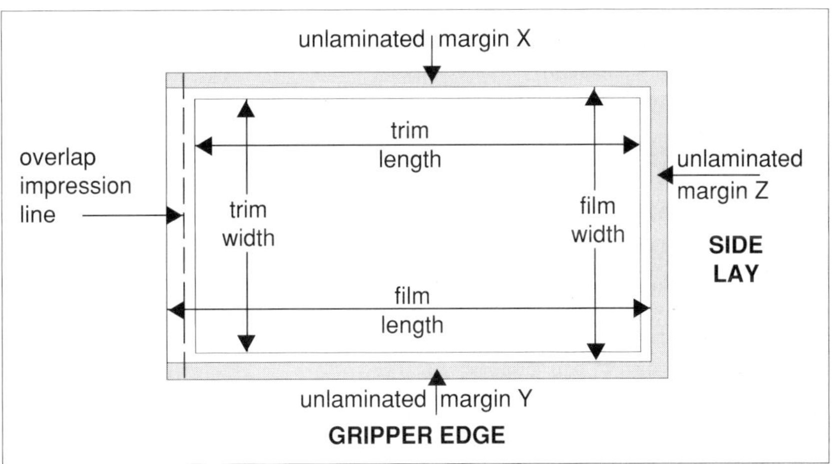

Figure 11.5: Margin allowances required for film lamination

Ideally, margins *X* and *Y* are *5mm each* and margin *Z 10mm*, but these may be modified according to particular circumstances. Generally, printers are happy to have approximately a *5 to 6mm coverage* over and above their final trimmed size.

3) Choice of stock
Although most types of paper and board can be film laminated, the best results are obtained from good-quality smooth material, either coated or uncoated. Very thin papers should be avoided, as they may become translucent. All sheets must be perfectly flat. Sheets which are cockled will be prone to creasing during the laminating operation.

4) Inks
It is generally found that a standard, good quality ink is suitable. Nevertheless, to ensure that no bleeding of ink pigments can occur, ink suppliers should be asked to supply inks with pigments which will normally resist the solvents used in laminating.

5) Anti-set-off spray
The presence of an excess of spray is always a possible danger to the final high-gloss result, resulting in a 'silvery' appearance, but provided a minimum of the finest spray particle is used and the spray apparatus is set to give an even distribution on the sheet, a good lamination can still be produced. If it is found that an excessive amount of anti-set-off powder spray has been used, the printed sheets will need to be hand dusted or run through a printing machine again, out of impression, to remove the unwanted excess powder.

Graining

Surface graining may be regarded as a means of converting the cheaper grades of paper into something more appealing or making normal grades look more interesting for special purposes. Any paper or board may be grained with, for example, a fine linen, canvas, hopsack, or pebble finish, the grain being chosen from a whole host of standard patterns. The process is applied to plain or printed sheets. It is not recommended to grain paper below 150g/m^2 if further processing is required: this is due to the sheet distortion which occurs during graining, creating problems with any final tight registration work carried out on the grained material.

In addition to adding a degree of refinement to the printed sheet, graining can also be used to mask minor defects in a paper of a printed result. Gloss or glare, where this is undesirable, may be masked by a suitable graining treatment.

Binding and finishing techniques

Saddle stitching

Design issues

When a double-page spread is to be used running across both pages, where possible this should be planned on the centre spread, which will avoid any problems with line-up across the two pages. Double-page spreads, if not reproduced on the centre spread, run the risk of misalignment across the pages, which is easily picked up by the eye - this problem is also experienced with other methods of binding using gathered sections such as thread sewn and perfect binding. It is especially difficult to avoid when thick, bulky stock is used.

As insetted saddle-stitched booklets or brochures become thicker the inside pages shorten, so much so that for every one millimetre in thickness, the page width reduces by approximately a corresponding one millimetre, eg an A4 portrait 3mm thick booklet will have centre pages measuring 207mm, not 210mm. When page margins are laid out this factor should be taken into account. If a repeat design is used, such as a solid or tint block which bleeds off the foredge, then some variation in the width of the printed area will result.

Production issues

To assist production, it is important that the grain direction of the paper should be parallel to the spine of the booklet wherever possible. Adequate head, tail and foredge trim margins should be allowed for, 3mm in thin booklets, but up to 5 and 7mm on thicker publications - these details should be checked out with the binder.

Covers on 135g/m^2 paper or heavier should be scored to avoid cracking; gloss coated stock of 135g/m^2 and above is particularly susceptible to cracking when folded, so pre-scoring is recommended.

On smaller booklets, the position of the wire stitches can be a problem. The allowance for every wire is 13mm and the minimum space between each wire 44mm, giving an overall length of 70mm.

Perfect binding

Design issues

Uncoated paper binds much better than coated paper and, for best results, the paper should not be too stiff or heavily calendered. The binding of landscape-bound books will cost more to produce as the majority of perfect binding machines will not collate and bind A4 landscape.

Ideally, printed illustrations and solid areas should not be printed right into the spine as inked areas adversely affect the adhesion properties of the binding.

It is advisable to have score marks 6mm from the spine, to allow the cover to bend cleanly on opening, and to avoid undue strain on the glued back cover area. The minimum thickness for the text of a bound publication is around 2mm; any thinner and the spine will bend under the pressure of the saw when the spine paper is being roughened off.

A sample dummy should always be made up using the correct materials, so that the printer and/or binder can calculate the cover spine width required and to check the general make-up of the publication.

Production issues

The recommended minimum trim allowance for a perfect bound book up to 6mm thick is 4mm, up to 10mm thick, 6mm, and for 10mm and over, 8mm; also the covers should, wherever possible, be supplied larger than the inside folded sections.

If single leaves and four-page sections have to be used, then they should be placed in the middle of the publication, ideally produced on paper of not less than 100 microns thick.

As far as size and thickness of bound copies are concerned, where collating and binding is done automatically the usual maximum is 380 x 255mm and minimum 155 x 110mm wide. Hand feeding and collating will allow for larger and smaller sizes than an automatic process. The thickness of a perfect bound publication can be between 2mm and 40mm, even greater on some specialist machines.

Great care should be taken to monitor the quantities produced for sections to be perfect bound, especially with the provision of overs. For orders up to 2000, overs required can be as high as 12% or more; up to 5000 or more, 10% or more. Covers, as they are usually subject to more processes than the text, should have an additional allowance to that for text sections.

Thread sewing

Design issues

Despite the high speed of modern sewing machine systems, it is more expensive and time consuming to thread sew a publication than to perfect bind by approximately 25% to 50%. Thread sewing is a high-quality method of binding which avoids the risk of leaves falling out, which can happen with perfect binding; it also opens easier and lies flatter when opened.

Thread sewing thin papers below 80g/m^2 should be avoided whenever possible and practical.

Printed areas that bleed and repeat throughout a publication should be kept away from the foredge as it is impossible to maintain a 'straight' look with thread sewn work.

Thread sewn jobs are produced quicker, and at lower cost, if the bound publication can be arranged so that it is produced with the same number of pages per section throughout - eg a 160 page book as 10 x 16pp sections. 'Odd' four-page or eight-page sections should be avoided as these will have to be placed around another section before sewing.

Production issues

The format size range of thread sewn publications is generally from a minimum of 150 x 115mm to a maximum of 350 x 250mm wide.

If possible, folded sections should have a lap of 6mm (known as a *binding lap*) allowed for in the back half of the section, which allows them to be machine fed faster and with lower spoilage - this working practice should be followed on all automatically fed bound work to gain higher speeds and lower spoilage.

Most print finishers prefer to sew in 16-page sections as eight-page sections are too thin; 32-page sections are possible, but heavyweight paper should not be used as creasing may occur at the head due to air being trapped, especially in the centre back margin.

Case binding

Design issues

If a paper cover and jacket are to be used, it is possible to utilise the same set of printing plates for both, achieving savings in time and cost.

Production issues

The substrate used for a paper case should be around 100g/m^2, with jacket and end papers around 135g/m^2. The end papers should be uncoated stock, similar in colour and texture to the text paper.

If gold blocking or blind embossing is required, binders blocks or zincos will need to be made to the requirements of each individual book.

Cover boards should always have the grain direction running parallel to the spine. For case bound books up to a text overall thickness of 10mm, 2.25mm cover board is usually satisfactory, with cover boards of 3mm for thicker books.

Folding

Design issues

It should be noted that the more folds that are used on a job, the greater the variation in subsequent fold positions; this is especially pronounced on thicker papers.

For double or open gate folded leaflets, the design should finish short in the centre, leaving a paper gap of 3mm to 15mm.

On work with multiple folds, it is important to take care that it does not become too thick or too small for the folding machine - as a guide, the normal thickness a fold unit will take is 2mm, giving a total folded thickness of 4mm. For multiple fold work, a paper that is resistant to cracking should be chosen.

If a concertina leaflet is required, allow more bleed margin on the foredge side of the cover page as this will need to be slightly wider to cover the folded leaves behind it.

Production issues

Work that is laminated or varnished both sides may not be able to be folded, although work with an aqueous coating applied on press is often suitable for folding.

Perfect bound sections that have to be folded should be perforated down the spine to release trapped air.

When endorse folding (folding the final finished product in half to aid handling, or to reduce the overall size to fit into an envelope or smaller package) is used, it is best to stagger any saddle stitch wires so that they do not meet other when folded in half. Nearly all endorse folded jobs will have paper creasing, the more folds and thicker carried out, the worse the creasing becomes.

Collaboration and consultation at all times on binding matters between the various production departments including design, and trade finishers if applicable, is essential.

This is to ensure any bound publication is finished as expeditiously, economically and efficiently as possible, and to the required standard.

This section on 'binding and finishing techniques' includes copy from the booklet *A guide to binding and finishing techniques for designers and printers* published by Jamesway Print Finishers.

12 Economical working practices and effective use of materials

Economical working practices, which include the most effective use of materials, are essential to all companies, large or small. Every organisation needs to pay close attention to ensure the methods they adopt and operate within their concerns are in line with best industry practice, otherwise they will not remain competitive, losing market share or at least not progressing as well as they could.

All production costs in their many forms, such as labour, equipment and materials, need to be well managed and controlled to ensure the organisation is utilising all resources as efficiently as possible.

Economies need to be sought in improving the form in which work is received from customers and its subsequent processing. It is essential, therefore, that there exists a framework of sound working practices and procedures which are followed to ensure each job is produced in the most efficient manner possible - in line with customer requirements plus the achievement of productivity and financial targets.

This chapter covers the following areas: keeping costs down and under control - economies effected in the preparatory and pre-press stages of jobs, economies in paper and board; factors affecting the working size of jobs; calculating the quantity of material required; allowances for wastage and overs; working out the cost of paper and board; makings and special makings; handling customers' paper; working to standard specifications; British paper sizes; printing machine sizes; deciding the paper and board to be used; different types of boards; paper specifications; printing processes and paper/board; and suitability of various substrates to a range of printed work.

The prices charged by printers to their customers are arrived at after the consideration of many factors, not least the current state of the economy in which they operate, the value of the job to the customer and degree of competition. This leads to the establishment of market prices, ie what a customer is prepared to pay for any particular job, based on current quotations and pricing from selected approved suppliers, taking into account previous pricing and service.

The amount of time taken on a job and the materials used, however, are also important considerations - these are the areas over which a printer has some control and therefore needs to ensure that efficient, effective procedures and sound working methods are practised.

It is primarily the function of the costing department to calculate and keep a close and up-to-date check on the cost of carrying out each process and operation in the production departments. It is, after all, of prime interest to any business that it should secure the fullest return from every form of expenditure, large or small, whether it be in time or materials.

The account executives should be knowledgeable about their company procedures and have an intelligent appreciation of what the relative costs of various operations are, since they will frequently become involved in determining the methods by which jobs will be produced.

The way in which instructions are given on the works instruction ticket - see *Chapter 2*, pages 25 to 28 - will clearly have a direct influence on the cost of production.

Keeping costs down and under control

Where no estimate has previously been prepared for a job, the account executive is required to ensure that the production costs are kept to an absolute minimum, consistent with the standard of quality which the work and customer demands. Whatever alternative methods or processes appear appropriate, they should always be considered seriously, and if account executives find themselves uncertain as to what are the most economic and practical solutions, they should not hesitate to seek the advice of estimating or production planning colleagues, or overseers.

The following suggestions and procedures examine some of the ways in which it is possible to ensure the most practical and cost-effective means of production. They are by no means exhaustive, neither are they all applicable to every job, but they should be studied with a view to establishing a framework which will benefit from the many possibilities, day-by-day, which will arise in processing printed matter through its many stages of production.

1 Economies effected in the preparatory stages of jobs

Checking the job requirements

Having the correct elements of the job prepared and available at the earliest possible stages is clearly the first step to saving time in its production. A considerable amount of time can be lost if the copy and/or instructions received from the customer or designer is not intelligible or marked-up with clear and sufficient detail.

If the customer provides very poor copy, in a mixed form, consisting of previously printed copy with marked-up alterations plus disks from different software programs and operating systems, then the account executive needs to explain tactfully the problems this will create.

It will be necessary to point out that the cost and additional time incurred in having the work prepared to a standard uniform format will be considerably less than planning all the different elements together, resulting in possible errors and inconsistencies.

Checking the space available for the printed matter

In order to avoid expensive reworking, it is a wise precaution whenever possible to check that all the production elements will fit into the areas indicated on the layout, before sending the job to the appropriate pre-press operation.

For the account executive to undertake this role on every job will be time-consuming, but it has been found in most printing organisations, however, that the time spent in job preparation in the preparatory stages is more than amply repaid in the savings which can be effected throughout all stages of production. To ensure the satisfactory completion of every job requires clear procedures and instructions, plus materials to the required quality and specification.

2 Economies effected in the pre-press stages of a job

Conventional reproduction

Whatever printing process is to be used, each classification of the elements to be reproduced by conventional means, such as bromides, line drawings, colour prints, other flat copy and transparencies, should wherever possible be grouped together to effect economy of camera work and subsequent planning.

Line work and continuous tone work will need to be treated separately. However, it must be remembered that when grouping, all elements must be of the same proportion of required size; colour prints or transparencies should be of the same colour density, otherwise additional time and cost will have to be expended to achieve the required result.

Artwork for reproduction of a bound publication can often be arranged in page form, or in double-page spreads, thus again reducing the time and cost of planning and assembly.

In the production of regular booklet or magazine work, layout grid sheets can be produced in advance: these indicate the page size, allowance for bled-off illustrations, the margins, width and lengths of columns, the exact position of the headings, folios and so on, all of which can be printed in light blue. These layout grid sheets provide an accurate guide for preparing the paste-up, thereby resulting in more cost-effective preparation of flat artwork, also known as the *mechanical*.

Digital reproduction

With the move towards more digital methods of working, as covered in *Chapter 7, Creating printed images*, it is essential that all parties in the print chain discuss and communicate their requirements at the earliest possible opportunity, so as to ensure a successful, compatible and as error-free as possible changeover from one stage to the next.

Customers, and designers working on their behalf, are producing an increasing amount of pre-press work on disk. The printer must therefore liaise with a wide range of contacts to ensure, as with conventional means of reproduction, that the form in which the work is supplied meets with the prime objectives of ease of access, speed and economy.

At some stage or other the printer will be responsible for giving or receiving instructions for the outputting of work supplied on disk.

Figure 12.1 takes the form of a disk outputting checklist which indicates the range of type of detail required by a printing company or bureau which provides an imagesetting service.

3 Economies in paper and board

Size. The printer should always use the smallest possible size of material, commensurate with technical requirements, which will accommodate the job to be done.

Due allowances for trimming, bleed and gripper - see page 198 - are obviously necessary, but the direct relationship between the finished size of the job and a standard size of printed paper and board is often overlooked by the designer or customer.

The account executive can often take the opportunity to make suitable suggestions to avoid unnecessary waste and save money. For example, reducing the finished, trimmed size of the job by just a fraction will often make it possible to produce the job economically on a standard stock size, instead of having to cut to waste.

Company details ______________________ Date ______________

Address ____

____

Contact name ______________________ Telephone no. __________

Date required ____

Essential information to be supplied

Job title ______________________ Purchase order no. __________

File name ______________________ No. of disks __________

Program used (including version number) ______________________

Imported graphics (names and generating programs with version numbers): __________

All disks should be labelled with the appropriate file names

Laser proof of all the work as it appears on the disks to be supplied and clearly identified

Page/format size: A4/A3/tabloid/other ______________ Reduction/enlargement to ___%

Crop registration mark: yes/no   Screen frequency ____ dpi   Screen orientation ___ degrees

Founts used (including those used in imported documents, specifying the manufacture of each fount):

____

Output information to be completed

Print the following pages: All ❑ from ____ to ____   Total no. of pages __________

Imposition software ❑ Spreads ❑ - details and instructions to be provided

Output to: bromide ❑ film ❑ negative ❑ positive ❑

Right-reading ❑ wrong-reading ❑ emulsion-side-up ❑ emulsion-side-down ❑

Resolution: 1270 dpi ❑ 2540 dpi ❑ other ❑

Separations: no separations ❑ all process colours ❑ all spot colours ❑

Special instructions

____

____

____

Figure 12.1: Disk outputting checklist for imagesetting

Substrate quality. Unless the paper or board has been specified by the customer or estimator, use the lightest weight consistent with good surface, colour and opacity: always consider existing stocks. The use of stock material often saves both time and money.

Self-covers. Sometimes it is possible to use the same paper for both the cover and the text pages - ie *self-cover.* This may save both machining time and the cost of material.

For certain types of work it may be found to be an overall economic advantage to use a slightly heavier weight of paper throughout, in order to achieve a self-cover job, rather than to incur the cost of printing the cover separately on a different paper on another machine.

This would avoid additional machine operations, also the need to insert the text into the covers. An obvious example would be a booklet comprising 12 pages and a four-page cover, which could be converted to 16 pages, self-cover - ie one sheet of paper, printed, folded and saddle-stitched as one section.

Select the right machine. Printing equipment and machinery, like any other production equipment, is highly specialised. Clearly, to utilise a high-speed automated colour press, intended to be used for printing large runs, for the production of say, jobbing work, such as letterheads, labels or office forms, in small quantities would be both inefficient and wasteful.

Impositions used. When producing bound publications check with the bindery the most suitable imposition scheme. To be able to use a folding pattern which can be done on in-house folding machines, rather than through a specialist trade house or by hand in-house can greatly affect the cost.

The material content of a print estimate can be as high as 50% depending on the complexity of the origination work, with paper and board normally accounting from a sixth to a third of the overall cost of a job.

It is therefore a major cost element in the calculation of an estimate and, as has been highlighted in recent years, paper and board are scarce resources. There are therefore environmental as well as economic pressures to use paper and board as efficiently as possible.

Factors affecting the working size of jobs

The decision regarding the paper or board to be used on jobs, rests at least partly on the suitability of quality and substance, which will be related to the printing process to be used, and economy of size.

The choice of quality and substance are often matters for discussion and approval, but the procedures and practices in establishing the most economical working sizes for jobs need special consideration in most cases.

The size of paper or board of the desired quality should be considered first. When a certain size is essential for economical working, it is advisable to base the specification on the nearest stock size which permits the working scheme that is required. The quantity may justify a making in a special size, if the time allows (see pages 204 and 205).

When selecting the size of substrate to be issued for working, the number of printed pages on the plate or other suitable image carrier must be considered, making suitable allowances for trim, bleeds, gripper, side and leave edges plus colour bar margins, and fold bulk in the backs where appropriate.

The amount of gripper space on sheet-fed machines will vary with different types of machine. If the printed matter comes close to the head and foot, a larger size of paper or board may be required to allow space for the gripper; alternatively, the possibility of gripping at the side may be considered.

Where work is to be printed work-and-tumble two gripper margins will be necessary. Work to be laminated or varnished sometimes requires a second gripper edge; often, however, the page margins may provide sufficient room for the grippers without the requirement for additional allowances.

The working size

The size of paper or board to be used and the method of working must be considered in relation to the quantity to be produced and the equipment available in the printing works - *see Chapter 8, Methods of working and impositions.*

Allowances for trim, grippers, registration and grain direction, sides and back leaves

Trim

An allowance for trim of 3mm should be made on all edges for single-leaf work, and on head, tail and foredge for bound work.

A special allowance, possibly up to 12mm, may be necessary for some subsequent operations, such as graining, varnishing, acetate lamination and for binding lap on automated binding machines.

For small forms and labels, special care should be taken in making the correct allowances, to ensure that the maximum number are printed on the sheet: some bled work such as sheet-fed labels, cards, leaflets and folders for quality work may require double cutting, as the outward portion of the substrate on the guillotine always shows a burred edge after cutting.

Jobs which are to be punched out, such as shaped labels, also require extra space, ranging from 3mm to 6mm between designs, but the use of dovetailing will result in economies - *dovetailing* is the butting together of irregular-shape items such as cartons and labels to gain the maximum number out of the material used.

In addition, an extra allowance of 6mm (3mm for each leaf) must be allowed for in the back margins of folded sections to be perfect bound.

The allowances indicated here may vary in some cases, and experience will soon build up a sufficient fund of reference information.

Gripper

An allowance for gripper margin on sheet-fed offset litho printing must be made on printed work which *bleeds* - (bleed is where the printed image is extended beyond the trimmed size without sufficient white margin space which can be used as the gripper margin). For small offset machines this can be as low as 6mm and for large machines it can be as high as 15mm.

Register and grain direction

The direction of the grain of the paper or board is very important in multi-colour close register work, especially when several press passes are required to complete the job - eg to complete a four-colour job on a single-colour machine will take four passes, two passes on a two-colour machine, etc.

To keep stretch to a minimum it as advisable to print *long-grain* - that is, with the grain running parallel to the impression cylinder. This is particularly important in sheet-fed offset litho.

Papers and boards should have the grain running along the impression cylinder, so that they lend themselves to bending round the cylinder.

If the grain is 'short', the tension of the paper or board may cause it to 'whip' and it may mark or crease and make register difficult to obtain. If the sheets need to be folded, the grain should ideally run parallel to the main folds to assist accuracy and to assist ease of folding, as well as to ensure correct opening in bookwork.

The grain direction for labels may be either vertical or horizontal to suit labelling machines and, therefore, customers' requirements need to be ascertained in advance of planning the most economical working method.

In carton work, the grain normally runs at right angles to the folds to suit carton gluing machines.

Most papers and boards are stocked long-grain. Paper mills and merchants will supply paper and board short-grain if required when available, or for a special making. Short-grain paper, if required, can be obtained by doubling the size of the stock paper obtained and cutting in half - eg SRA1 (640mm x 900mm) long-grain becomes SRA2 (450mm x 640mm) short-grain.

Long and short-grain paper or board is identified by the use of a recognised and uniform means of description - eg long-grain SRA2 is described as 450 x 640(m)mm and short-grain as 450(m)mm x 640mm - 'm' indicating the direction of the grain.

Sides and back leave margins

When printing by sheet-fed offset litho an allowance of 6mm on the two side margins and 6mm on the back leave of the sheet can be made to ensure damp/water will not be visible on the outside edges of the printed sheet when trimmed to final size. As a gripper allowance has already been made, there is no need to make a further allowance on the front leaf of the sheet.

To help illustrate the use of trim allowances, these four examples of jobs have different requirements:

i) illustrates a poster printed singly with 3mm trim all around

ii) illustrates an eight-page booklet with 3mm trim as required

iii) illustrates a label printed eight-up (multiple images) with *single cuts* through the multiple images and 3mm trim allowance only on the outer edges

iv) again illustrates a label printed eight-up (multiple images), but this time with *double cuts* for higher quality work.

As previously stated, guillotine cutting and trimming produces for every cut of the knife, one clean edge and one 'burred' edge - this would be the result for the labels in example (iii). To avoid a burred edge on the labels it is necessary to allow for double cuts as in example (iv).

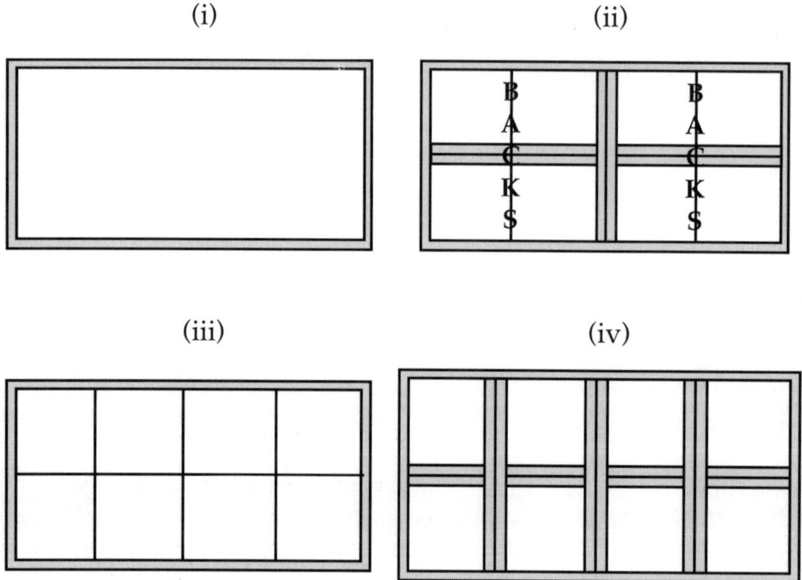

The thin shaded area indicates 3mm trim

In order to reinforce learning and understanding of this area, it is helpful to express the illustrations in the form of structured calculations for the four examples, with a finished trimmed size of A2 for example (i) and A5 for examples (ii), (iii) and (iv).

(i) *mm*		*mm*		*(ii)* *mm*		*mm*
420	x	594	**trimmed size**	210	x	148
6		6	trim	6		3
				216	x	151
			8 pages to view	2		4
426	**x**	**600**	**working size**	**432**	**x**	**604**

(iii) *mm*		*mm*		*(iv)* *mm*		*mm*
210	x	148	**trimmed size**	210	x	148
			trim	6		6
				216	x	154
2		4	8-up	2		4
420	x	592				
6		6	trim			
426	**x**	**598**	**working size**	**432**	**x**	**616**

The following example illustrates an eight-page, A4 trimmed to bleed, insetted section, printed in four-colour process.

mm		*mm*	
297		210	**trimmed size**
6		3	trim
303	x	213	
2		4	8 pages to view
606	x	852	
6		12	sides and leave edges
10			colour bar
12			gripper margin
634	**x**	**864**	**working size**

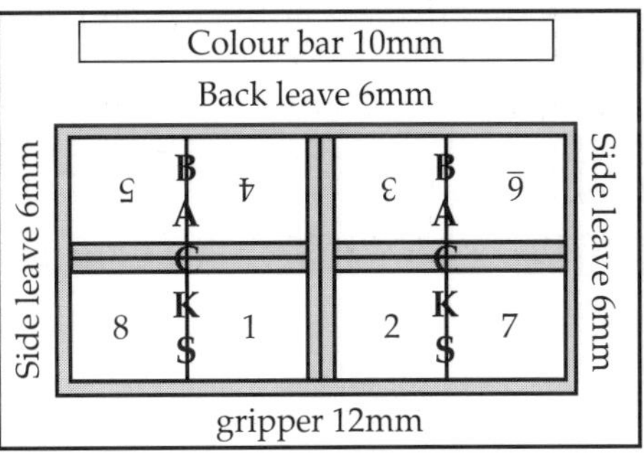

Nearest stock size SRA1 - 640 x 900mm

Calculating the quantity of material required

The number of sheets, or cylinder circumference/cut-offs on reel-fed work, or parts thereof, required for each finished copy is multiplied by the number of copies to be printed, so as to find out the total quantity of material required for each job.

Care must be taken not to calculate half or double the quantity required by confusing leaves with pages, and to remember that there are two copies out when printing one-set work-and-turn, also on reel-fed work, where variable-sized cylinders are used, to take account of the correct number of copies out, across and around the cylinder.

Examples

Magazines

20 000 copies, consisting of 68 pages self-cover, trimmed size A4
Worked on SRA1 paper size
SRA1 for 8 plates of 8 pages = 4.0 sheets
1 plate of 4 pages = .25 sheet

20 000 copies x 4.25 sheets = *85 000 sheets of SRA1, excluding allowance for wastage, required for this job.*

Jobbing

16 000 leaflets, trimmed size A5, printed both sides
Worked four-set, work-and-turn, on SRA2 paper size

16 000 ÷ 8 out = *2000 boards of SRA2, excluding allowance for wastage, required for this job.*

Bookwork

5000 copies, consisting of 256 pages metric crown 8vo
Worked on metric quad crown paper
2 plates x 32 pages sheet-work = 64 pages to a full sheet
256 ÷ 64 = 4 sheets x 5000 = *20 000 sheets of metric quad crown, excluding allowance for wastage, required for this job.*

Continuous business forms

30 000 single-part forms, finished size A4
Worked on cylinder size 24.75" with 26" web width - see *Figure 8.14*, page 134
30 000 ÷ 6 out = *5000 full cylinder size, excluding allowance for wastage, required for this job.*

Allowances for wastage and overs

Overs must be added to the basic quantity of paper and board required for the job to allow for proofs, spoilage, and house files.

The quantity, inclusive of wastage and overs, will depend upon the number of colours or workings, class of work, closeness of register, class of substrate, type of machine and length of run, plus additional allowances for any binding and outwork processes.

On short runs, a higher proportion of overs is necessary, as a number of sheets or web lengths are wasted in the preliminary stages of machining, whatever the length of the run.

In multi-colour work, the allowance must be sufficient to cover such risks as bad register, defective colour and marking. If in doubt allow a little extra, since it is important to avoid expensive reprints.

Allowances for wastage and methods of calculation vary from company to company, but the following formulae might be used as a starting point:

Suggested formulae for calculating litho machine and bindery wastage:

a) Sheet-fed single- and two-colour presses

200 sheets for each make-ready + 5% overs for each pass through press which includes overs for standard finishing.

b) Sheet-fed four-colour presses

300 sheets for each make-ready + 6% overs for each pass through press which includes overs for standard finishing.

c) Reel-fed single-colour

2000 impressions make-ready + 8% overs.

d) Web offset multi-colour

3500 impressions make-ready + 10% overs.

Note, for each additional web, an extra make-ready allowance of between 25% and 50% of the impressions allowed for the first web should be made.

Working out the cost of paper and board

When the total number of sheets or weight in reels, including provision for overs, has been ascertained, and the relevant prices are known, the calculation of cost is a relatively simple matter.

Examples

Calculating the cost of 9400 sheets of matt-coated cartridge, size 505mm x 770mm, weighing 25.3kg per 1000 sheets (65g/m^2) at £653 per tonne.
The cost of 1000 sheets works out at 25. 3 x 65.3p - ie £16.52, therefore, *9400 sheets will cost £155.29.*

Calculating the cost of 1750 boards of white pulp, size 570mm x 730mm, weighing 9.6kg per 100 boards (230g/m^2) at £680 per tonne. The cost of 100 boards works out at 9.6 x 68.0p - ie £6.53, therefore, *1750 boards will cost £114.28.*

Calculating the cost of 19.14 tonnes of blade-coated 60g/m^2 at £650 per tonne, supplied in reels for heat-set web-offset.
The total cost works out at 19.14 x £650 = £12 441.

Makings and special makings

A *making* is a non-standard size in a standard grammage and material - for example, 690 x 870mm in 100g/m^2 matt-coated.

A *special making* is a non-standard size in a non-standard grammage and/or material - for example, 690 x 870mm in 145g/m^2 matt-coated.

The minimum quantity for a making is usually not less than two tonnes for ordinary papers; special making orders can be as high as 50 tonnes. In either case, the mill should be contacted for minimum tonnage and delivery dates.

In a world where printers' customers are demanding exact quantities of print (with no unders or overs) and papermakers' technical control of variables has increased enormously, printers are demanding making tolerances far tighter than those laid down in British Paper and Board Trade Customs, 1988. This is especially true of making quantities. While the theoretical position which some mills will resort to is given in the table overleaf, many printers are looking on large orders, for +/- 1% or, in practical terms 'no unders, with overs restricted to the nearest reel or pallet'.

Buyers and makers of paper and board are at liberty to agree specific conditions of sale (including making quantities) for individual orders.

	Makings	**Special makings**
	Standard stock quality and grammage in special sizes	*Non-standard papers - eg by reasons of quality and/or grammage*
	%	%
Up to and including 1 tonne	10	15
Over 1 tonne and not exceeding 5 tonnes	5	10
Over 5 tonnes and not exceeding 10 tonnes	5	7.5
Over 10 tonnes and not exceeding 20 tonnes	3.75	5
Over 20 tonnes	2.5	2.5

Breakage charges

Where the quantity of paper or board required for a job is less than that normally packed in a parcel - eg paper in 500 sheets or board in 100s, it should be noted that the paper supplier will only break parcels of unusual stocks. In such instances, there is an additional cost for handling in splitting and repacking known as a *breakage charge.*

Handling customers' paper

In certain classes of work, such as periodicals and bookwork, the paper and board used may amount to large tonnages per annum. These are often ordered by the publisher as makings and special makings, and delivered direct to the printer. Records must be kept by the printer of all receipts and issues, and a notification sent to the publisher, showing the stock balances, after every issue has been produced.

Where the paper or board is to be supplied by the customer, it is advisable to arrange for *out-turn sheets* to be supplied to, and checked by, the printer before the paper or board is dispatched by the customer, paper merchant or mill, in case the material is found to be unsuitable for any reason.

When paper or board has been held in stock for a long period, it should be examined thoroughly upon receipt. Seriously damaged edges, or the effect of dampness, for example, could cause a great deal of trouble later. Such matters must always be taken up with the customer without delay.

Working to standard specifications

Paper and board is supplied to printers in sheets or reels, depending on the press for which it is intended. Many printed products are available in standard sizes and printing equipment reflects this fact. Paper on the reel or in sheets is supplied to dimensions which will accommodate such sizes.

For centuries each country had a large range of untrimmed sizes - some of which were common to all grades while others applied only to one kind of paper - for instance, a printing or writing paper.

Most papers are now supplied to printers for conversion into standard product sizes based on the International Standards Organisation 'A' range of sizes covered by BS 4000 Part 1: 1990. These were intended to replace the existing British sizes over a period of time. In some cases, rather than replacing them, they have become an addition to the stock range. This is likely to continue for some time to come. In particular, the business forms market continues to manufacture forms based on inch sizes (even if they are quoted in metric sizes) as the computer printer market is dominated by US and Japanese suppliers, who still use the inch as their means of measurement.

Many of the traditional printing and book sheet sizes in the UK were changed slightly on metrication when the finished sizes of books were also changed. These were then given metric sizes to the nearest mm. The printing and paper industries were pioneers in UK-metrication, but they soon found that the continental paper mills were centimetric rather than millimetric and did not want to cut sheets a few mm larger or smaller than those for their home markets. As a result, some standard sizes can be found in the original mm sizes as well as in dimensions rounded up or down to the nearest cm or half cm.

International paper sizes

There are three interrelated standard ranges for paper and board: each range within the series being designated by an initial - *A, B* or *C*. Most printers are concerned only with the A range but, in between each of the sizes in the A range, comes those of the B range, filling any gaps and intended primarily for poster work. The C range consists of the finished sizes of envelopes and folders which will contain A-sized items. All the items in the ISO series are proportionate to each other, as shown by *Figure 12.2*.

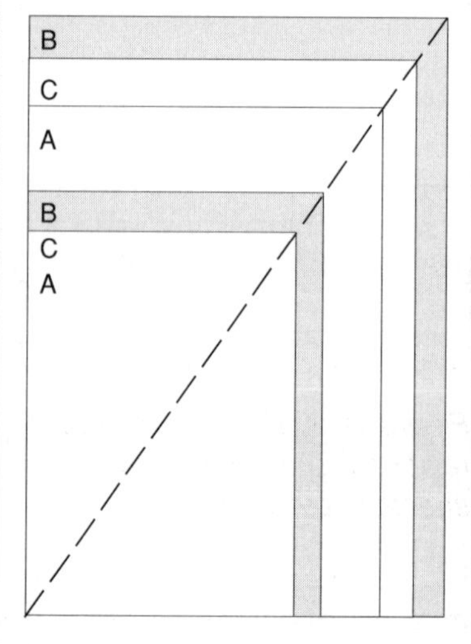

Figure 12.2: All items in the ISO series are in proportion

The basis of the international series of paper sizes is a rectangle having an area of one square metre (1 000 000mm^2), the sides of which are in the ratio of 1: √2 (or 1:1.414). This basic size is A0 or 841 x 1189mm, the geometrical relationship of the sides being shown in *Figure 12.3*. The ratio of the sides has the unique property of being retained when the longer side is halved or the shorter side doubled. This makes it attractive to the graphic designer and to anyone who has to reproduce their work in different standard sizes.

Multiples and sub-divisions

The basic size in each range is either *A0, B0* or *C0*. If this is preceded by a figure - for example, 2A0, it indicates that the area of the basic A0 sheet has been doubled by doubling the shorter dimension. If the letter designating the series is followed by a figure, for example A1, it indicates that the area of the basic A0 sheet has been halved by halving the longer dimension. Similarly, A2 is half A1, A3 is half A2, A4 is half A3 and so on *(Figure 12.4)*.

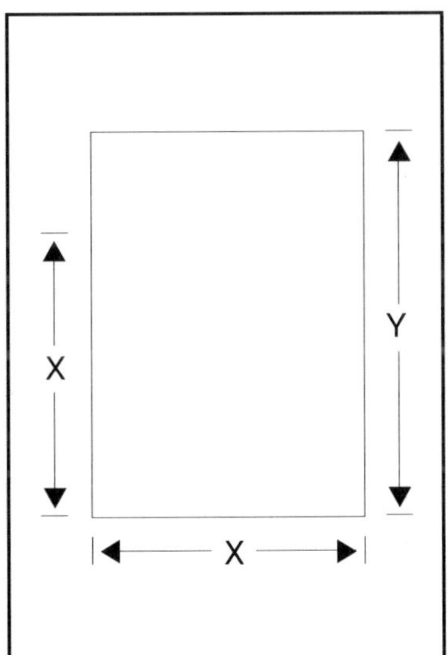

Figure 12.3: The geometrical relationship of the sides of sheets in an A size

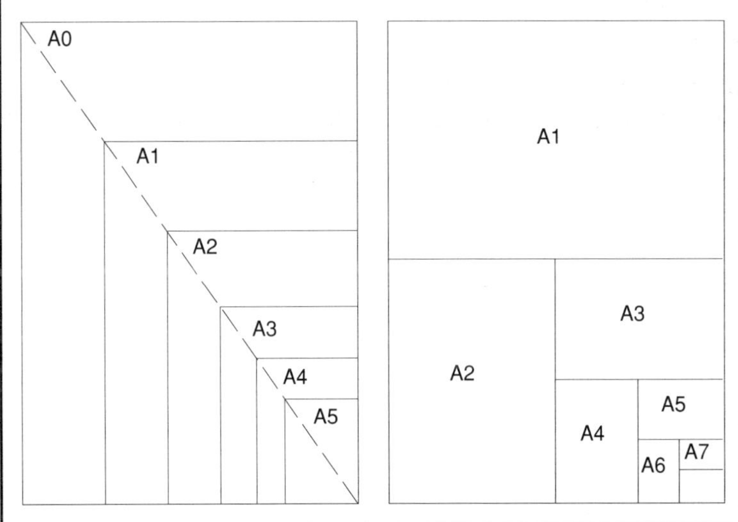

Figure 12.4: The area of the basic A0 sheet is halved by halving the longer dimension

While there are exceptions (such as books) A4 and A5 have become the most popular sizes for the majority of printed reading matter, replacing quarto and foolscap in many areas. The major exception is in North America where the American version of quarto remains. *Figure 12.5* overleaf shows the multiples and subdivisions for each range.

A range mm		B range mm		C range mm	
2A0	1189 x 1682	2B0	1414 x 2000		
A0	841 x 1189	B0	1000 x 1414		
A1	594 x 841	B1	707 x 1000		
A2	420 x 594	B2	500 x 707		
A3	297 x 420	B3	353 x 500		
A4	210 x 297	B4	250 x 353	C4	229 x 324
A5	148 x 210	B5	176 x 176	C6	114 x 162
A7	74 x 105	B7	88 x 125	C7	81 x 114
A8	52 x 74	B8	62 x 88	C8	57 x 81

Figure 12.5: Multiples and subdivisions for A, B and C ranges

Untrimmed stock sizes of paper

The object of introducing ISO sizes to the UK was to reduce the number of sizes stocked and to introduce the concept of a finished-product standard size as opposed to a standard printers' raw-sheet size. Two untrimmed stock size ranges have been introduced - the *RA* and *SRA* ranges - to produce A-size finished products. The RA sub-range is not widely stocked. Most stockists hold the SRA sub-range while, for writing and reprographic papers, RA sizes are held. B sizes are much less used, though B4 is common in reprographics as it is approximately the size of standard computer print-out. The proportions of A0, RA0 and SRA0 are shown in *Figure 12.6.*

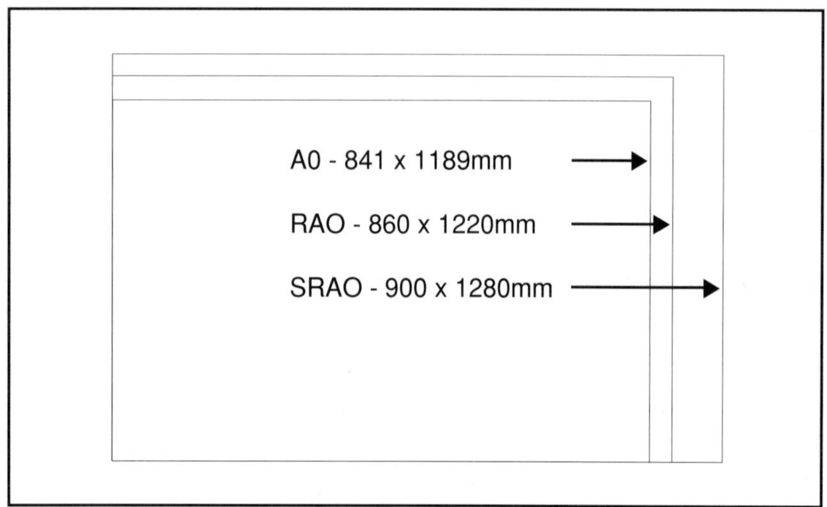

Figure 12.6: Proportions for A0, RA0 and SRA0 ranges

Primary range mm	Supplementary range mm
RA0 860 x 1220 RA1 610 x 860 RA2 430 x 610	SRA0 900 x 1280 SRA1 640 x 900 SRA2 450 x 640

Figure12.7: Multiples and subdivisions for RA and SRA ranges

The primary range (*allowing 5% waste*) is suitable for work in which a normal trim is required. Most jobs are trimmed at some stage of being finished and, as explained previously in this chapter, the normal trim is 3mm. After trimming, a standard size job results. For example, an eight-page A5 upright booklet, trimmed size 210 x 148mm can be printed on an RA2 sheet (430 x 610mm), if available in this stock size.

The untrimmed size, after folding and stitching, is 215 x 152.5mm - 3mm would be trimmed off the head and 2mm off the tail to reduce the 215mm to 210mm and 4.5mm would be trimmed off the fore-edge to reduce the 152.5mm to 148mm.

If, however, the work prints close to the edge of the sheet, a larger untrimmed sheet is necessary. The same job, trimmed to bleed, could be printed on an SRA2 sheet (450 x 640mm) from the supplementary range. After folding and stitching, the untrimmed size would be 225 x 160mm. A 6mm head trim and a 9mm tail trim reduces the 225mm to 210mm, while a 12mm fore-edge trim reduces the 160mm to 148mm. *The SRA range allows 15% waste above the corresponding A size.*

Apart from the RA and SRA range and some old imperial British paper sizes still retained mainly for bookwork, paper mills and merchants now offer, in addition, a range of stock sizes to cope with a demand for a wider range and to suit differing pagination formats.

These include:

- 335 x 640mm and 340 x 640mm - to suit six pages A4 or 12 pages A5

- 520 x 720mm - B2 plus, following the growth in printing machines of this size

- 640 x 650mm - to suit six pages A4

- 630 x 880mm - saving on paper compared to SRA1 (640 x 900mm) when working size allows

- 720 x 1020mm - B1 plus, again following the growth of printing machines of this size.

British standard paper and board sizes

British standard paper dimensions have always been stated in untrimmed sizes. Unlike international sizes, they are known by name and there is no proportional relationship between the various sizes - see *Figure 12.8*. Similarly, multiples and sub-divisions are named, whereas in the ISO system they are numbered.

Another anomaly is that it is possible to have two different finished sizes - for example, metric crown 8vo can be either 186 x 123mm or 180 x 120mm, depending on whether the job has normal trims or is trimmed to bleed, and it is likely to vary between landscape and portrait versions as a landscape booklet may need a greater final head trim. Centimetric versions are often stocked. *Figure 12.9* lists a range of the most popular British standard paper and board sizes.

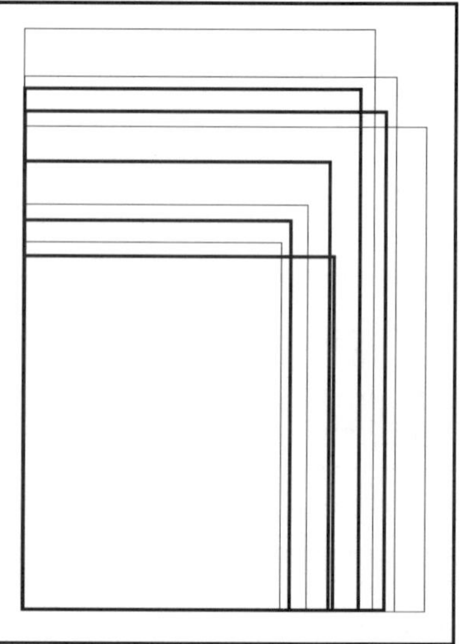

Figure 12.8: Illustration confirming there is no proportional relationship between traditional British paper sizes

	mm	mm also centrimetric
Printings		
Metric double crown	505 x 777	(520 x 760, 510 x 760, 505 x 770, 510 x 710)
Metric double demy	564 x 888	
Double medium	584 x 914	(585 x 915)
Double royal	636 x 1016	(635 x 1020, 640 x 1020)
Writings		
Large post	419 x 533	(420 x 535, 425 x 535)
Medium	457 x 584	
Double cap	432 x 686	(430x 685, 430 x 690)
Boards		
Royal	521 x 636	(520 x 640, 520 x 6350
Postal	572 x 724	(570 x 730, 570 x 725)
Imperial	572 x 786	
Index	648 x 762	(650 x 775)

Figure 12.9: British standard paper and board sizes

Sub-divisions of British paper sizes

A piece of paper, when cut to size, is described as a single leaf which consists of two pages - one page on each side of the leaf. Thus a book made up of 100 single folded pieces of paper consists of 100 leaves or 200 pages - the number of pages being double the number of leaves. Whatever its size, every sheet of paper has a number of sub-divisions which are common to all sizes of paper.

In its basic size, a sheet is known as *broadside*. When folded midway along its longer dimensions to give two equal leaves or four pages, it is called a sheet in *folio*. If the folio sheet is again folded in the middle of its longer dimension, it will then be a quarter of the size of the full sheet, consisting of four leaves or eight pages and is called *quarto (4to)*. Further folds will produce *octavo (8vo)* and so on. Each fold halves the superficial area and doubles the number of leaves or pages obtained by the previous fold. The various names are used not only when sheets are being folded but also when being cut.

Other less common sub-divisions are 6mo when the long dimension is divided by three and the short dimension by two, to give six leaves or 12 pages, and ordinary 12mo when the long dimension is divided by three and the short dimension by four to give 12 leaves or 24 pages. *Figure 12.10* shows metric crown sub-divisions in untrimmed mm.

metric crown sub-divisions	mm (untrimmed)
broadside	384 x 504
folio	252 x 384
4to	192 x 252
8vo	126 x 192
16mo	96 x 126
32mo	63 x 96
6mo	168 x 192
12mo	96 x 168

Figure 12.10: Examples of sub-divisions for metric crown

For orders of several tonnes, depending on each mill's minimum quantity, the printer can order a making of any size. The only limit may well be the width of the wire (or deckle) on the papermaking machine. Special sizes may also be accommodated by cutting from mill reels. Some merchants and mills also stock reels for mini- or standard-web presses in a restricted range of qualities. Reels, more often than sheets, are ordered as makings as the tonnages are frequently higher.

Printing machine sizes

Each printing process has available to it a range of printing machinery, in sheet and web format, single and multi-unit, which has been made in response to the type of printed work and products to which the equipment is best suited to in terms of quality and cost-effectiveness.

It is through being aware of the equipment available and its full potential within a company that we are able to calculate the most efficient and practical methods of working best suited to it. It is a matter of priority that the account executive has a copy of the current company plant list and is fully conversant with all the printing machines in terms of minimum and maximum sizes, number of units, quality limitations, in-line finishing potential and appropriate ancillary equipment.

Chapter 10, Printing processes, pages 161 to 163 covered machine printing running speeds and the range of available printing machines which can be found in common use in printing companies.

Offset litho has the largest variety of machines available, in terms of both of sheet- and web-fed. To assist in the understanding of this area the most common range of both sheet-fed and heat-set web offset presses will be considered.

Sheet-fed offset litho printing machine sizes

Model	*maximum sheet size*
***00**	**360 x 520mm**
01	480 x 660mm
***0B**	**520 x 720mm**
2C	640 x 915mm
***3B**	**720 x 1020mm**
5	890 x 1260mm
5W	920 x 1300mm
6	1000 x 1400mm
7	1100 x 1600mm
7B	1200 x 1600mm

The above is a range of press model sizes offered by a wide range of printing machine manufacturers. The list is by no means exhaustive, although it does cover the main models used by printers. The maximum sheet size as indicated may differ slightly between different manufacturers.

***00 - 360 x 560mm** (equivalent to B3) allows 2 x A4 to-view with extra for bleeds, etc. This is the most popular small offset size, with presses available in mainly one, two, four and five colours. There are some press sizes below this size which have a maximum sheet size of 280 x 394mm.

01 - 480 x 660mm is popular as an SRA2+ size press printing up to 4 x A4 to-view with extra for bleeds, etc. Presses are available in mainly one, two, four, five and six colours.

***0B - 520 x 720mm** (equivalent to B2) is now tending to replace the 01 press size. Again it allows 4 x A4 to-view with considerable extra room to allow for bleeds, 'oversize' products, cover flaps, etc. Presses again are available mainly in one to six colours.

2C - 640 x 915mm (in some ways this takes the form of the larger partner to 01) is an SRA1+ size press printing up to 8 x A4 to-view with extra for bleeds, etc. Presses are normally only available in one, two and four colours, although five colours have been made to order.

***3B - 720 x 1020mm** (equivalent to B1) Together with the 0B models these are the most popular sizes for general commercial printers. Prints 8 x A4 to-view with plenty of extra to spare. Presses are available in one, two, four, five, six, seven and eight colours.

5 - 890 x 1260mm and
5W - 920 x 1300mm - These two sizes are used by specialist producers such as carton or book printers. Presses are available mainly in two, four, five and six colours.

6 - 1000 x 1400mm (very close to final trimmed size B0 - 1000 x 1414mm)
7 - 1100 x 1600mm
7B - 1200 x 1600mm - These three sizes represent extra-large presses capable of printing 16 x A4 to-view with plenty to spare. Popular with carton printers. Available mainly in two, four, six and eight colours.

Multi-colour convertible printing presses are now becoming popular as well as perfecting presses, eg a two-colour convertible can either print 2:0 or 1:1 - ie two colours on one side only or one colour both sides in one press pass: the most common configuration of a five-colour convertible is 5:0 or 4:1 - ie five colours on one side only or four colours on one side and one colour on reverse. This facility is of tremendous benefit to printing companies as it allows machines to be much more flexible and productive.

A 'true' *perfector* always prints on both sides of the sheet at the same time - ie 1:1, 2:2, 4:4 - one-back-one, two-back-two, or four-back-four colours in one press pass.

Heat-set web offset litho printing machine sizes

The majority of large format web offset machines built for the UK market have a *cut-off or cylinder circumference of 630mm,* although in recent years 625mm and 620mm cut-offs have been introduced to save paper; some very large machines also have a double circumference cut-off of 1260mm.

Cylinder web widths vary up to 965mm wide, although some *short grain presses* can be wider. Traditionally, printing cylinders are greater in width size than they are in circumference, whereas short grain presses reverse this situation with the width of the cylinder being greater than the circumference.

Configurations of presses vary, but the following simplified list gives some of the variations available in heat-set web offset presses. Presses are identified by the number of A4 pages out from one cylinder revolution.

	No. of webs	Web width	Circumference/cut off
8 pages 'mini' web	1	508mm	630mm
16 pages	1	965mm	630mm
32 pages	2	965mm	630mm
32 pages	1	965mm	1260mm (630 x 2)
64 pages	2	965mm	1260mm (630 x 2)

Figure 12.11: Range of the most popular heat-set web offset press configurations

Deciding the paper and board to be used

If a job has already been the subject of an estimate, the quality and weight of the paper or board to be used will have been determined by the customer or estimating department.

It may simply be a straightforward reprint of an item which has been produced before, either in your own works or by another printer. On the other hand, particularly if it is an entirely new job, the customer may have supplied a previously printed sample with the request that the substrate for the job should be of a similar quality, or he may not have given any indication at all, since he cannot judge what is most suitable for it.

Three main factors must come under consideration when deciding upon the substrate to be used:

i) the most suitable quality, size and weight of paper or board to be used

ii) the comparative cost factors between alternative qualities

iii) the availability of the quality required, either from the printer's own stock, paper merchant or mill.

With the multitude of different grades of paper and board available today there is at least one quality, and in many cases several qualities, suitable for every job, every purpose and every printing process - knowledge of a wide range of paper and board grades and their general suitability is therefore essential.

Varieties of paper

Whilst basically the principles of papermaking are the same for all kinds of machine-made papers, the characteristics required in a paper or board vary according to the purpose for which it is used. These variations are obtained by the use of different furnishes and by differences in the processes of making the substrates.

As a general rule, paper mills are laid out for handling certain kinds of material - one mill may produce papers from predominantly recycled stock, another from mechanical wood or chemical wood, so consequently papers made from different pulps generally have to be made at different mills. In some cases a mill may specialise in one kind of paper, for instance, coated paper, whilst another mill may produce a range of different kinds or printing and writing papers, all produced from the same main raw materials.

Most kinds of paper and boards are produced in various qualities depending on the furnish and vary in price accordingly. Mechanical papers are the cheapest and rag papers the most expensive, esparto papers being considerably cheaper than rag papers but slightly more expensive than wood-free papers.

Newsprint

Newsprint is a mechanical pulp printing paper containing a small proportion of chemical wood pulp. It is used for newspapers and newspaper-type products - for example, some tabloid house journals - usually printed on rotary presses by cold-set web offset, flexography and letterpress. Being unsized, it has the property of absorbing the relatively liquid inks used on newspaper presses, but quickly discolours on exposure to daylight and sunlight due to the high percentage of mechanical pulp in its furnish. It is the cheapest grade of printing paper and is supplied in reels or sheets, machine-finished for cold-set web offset and letterpress. Grammage range normally covers from 36 to $50g/m^2$.

Mechanical printing

This is a superior type of newsprint containing a larger percentage of chemical wood pulp, which varies according to the quality required. Better qualities have such a small percentage of mechanical wood that the appearance of the sheet is almost that of a woodfree printing. Containing only an ordinary amount of size, this class of paper is generally used for cheap publications and similar work, but it may also be hard-sized and used as cheap writing paper. The presence of mechanical wood causes discoloration and coloured mechanical printings frequently have a mottled appearance through the failure of the mixed furnish to take the dyestuffs uniformly. It is supplied in reels or sheets.

Mechanical sc printing

Having high loading and a smooth gloss surface obtained by a supercalender, mechanical sc printing is used extensively by rotary gravure for products such as mass circulation magazines, colour supplements and catalogues, in a grammage range from approximately 54 to $80g/m^2$. Similar paper, prepared primarily for heat-set web offset, is gum-surfaced sized, as are bulky mechanicals prepared for cheaper books.

Woodfree printings

Woodfree printings (that is, free from mechanical wood pulp) are made from chemical wood pulp, which produces a clean sheet of good colour, suitable for all kinds of general printing and magazine work, and stronger than mechanical printing. Supplied machine-finished (mf) or supercalendered and in reels or sheets, they are known as litho printings. Woodfree printings are used for a wide range of good quality commercial printing work including leaflets, booklets, reports and books, in an extensive grammage range from approximately 60 to 135g/m^2.

Bible paper

Bible paper is an extremely thin, white opaque printing and is so named because it is produced for bibles and other work for which a large amount of reading matter has to be condensed into a book of modest proportions and high pagination - for which the paper must be both thin and opaque. Heavily loaded, it reproduces halftone screens, but is unsuitable for writing. Most grades are made from chemical wood, but the best grades have a cotton or linen finish.

Woodfree antique wove

The term antique was originally used to describe papers made by machine in imitation of old hand-made papers but is now used to describe any paper with a good bulk and a rough surface. Antiques are light in texture and their use in book work results in a thicker book than when mf or sc paper of similar grammage is used, as the sheet is thicker relative to weight.

Antique wove

Antique wove is less bulky than antique laid and usually presents a better printing surface due to the greater and more even compression produced by the closely woven dandy roll on the paper machine, but neither wove nor laid is suitable for anything but the coarsest halftone printing, both having a rough surface. Grammage range varies from approximately 70 to 90g/m^2.

Cartridge

Cartridge is similar in appearance to hard antique paper, a wide range of cartridges being produced for printing as well as drawing. A tough paper usually made from wood pulp and some of short fibre, grammage is from 60 to 170g/m^2 and higher. The amount of sizing is dependent on the purpose for which it is to be used and its surface may be rough, semi-rough, uncoated, or coated.

Offset printings

Offset printings (offset cartridge is now a misnomer as it covers so many papers) are specific printing papers, cheaper than drawing cartridge, produced for litho printing and usually somewhat whiter than the creamy shade of traditional cartridges. Its hard-sized and surface-sized finish, free from fluff, makes it particularly suitable for offset litho - hence its name. The best grades of offset papers are made on twin-wire machines.

Mg poster

Mg poster papers have a smooth surface on one side and a rough finish on the other and are described as *machine glazed* or *mill glazed (mg)*. Although there are many kinds of mg papers produced for use as wrapping papers and for other purposes, those most commonly used for printing are known as *posters*, being printed on the smooth side and pasted on the rough side for sticking to the hoarding. Mg papers are made on special machines and are hard-sized to resist the penetration of paste. A variety of white and tinted qualities are produced, such as mg mechanical poster, mg woodfree poster and mg litho poster. Grammage range covers from 70 to 170g/m^2.

Coated papers

Art paper is a term used to describe the best quality of coated papers, having high coating weight, special pigments and a very smooth surface, usually with a high gloss. The best art paper finishes are obtained by using a twin-wire base and applying two coatings of china clay or other mineral to each side as a separate operation after the body paper has been made. Different grades of coated paper are distinguished not only by the quality of the body paper, but also by the quality of the coating material, and the amount of calendering to which it is subjected in the finishing process. Most papers using a woodfree base and of a high quality are known as *gloss-coated*.

To meet the demands for colourful glossy publications, the paper industry has concentrated on the production of low-priced coated paper, suitable for litho printing, using a part-mechanical body.

Methods of applying coating to the body paper may be an *on-machine* process where the coating equipment is an integral part of the papermaking machine, or an off-machine method where the coating is applied on a machine entirely separate from the papermaking machine.

The coating mixture is applied by means of rollers or blades, which meet the paper web and transfer the coating to the web, where its thickness and smoothness is controlled either by smoothing rollers, a steel blade or an air knife, that is, a high-pressure jet of air from very fine apertures across the web of paper. These methods are sometimes used in conjunction with each other.

As with ordinary papers, coated papers may be calendered on the papermaking machine or supercalendered on a separate machine, depending upon the intended use and the degree of finish required. Qualities vary with the furnish of the base paper, which may be mechanical or woodfree. Machine-coated papers are used mainly for magazine and other illustrated work requiring better halftone reproduction than is possible on mf or sc paper, but for which the better quality coated paper is too expensive.

Matt-coated papers

Matt-coated papers have become popular in recent years due to the lack of glare, often associated with gloss papers, and to the soft feel and texture to the paper. Depending on the finishing process and/or suppliers' descriptions, papers may be termed *matt, satin, velvet* or *silk*, also giving some indication as to the coating's smoothness.

A matt-coated finish is achieved by using a considerable amount of calcium carbonate (chalk) in the papermaking process. The printability and smoothness of the paper is determined firstly by the amount of coating applied to the base paper surface and, secondly, the finish.

Unfortunately, although matt-coated papers are very popular, they can provide problems for the printer in terms of the *sandpaper effect* with poor rub-resistant properties.

Matt-coated papers appear to be smooth, but in fact are quite rough when compared with gloss-coated, so producing the sandpaper effect when two printed surfaces come together. Ink rub and marking can occur on printed surfaces long after the ink has fully dried. Specially formulated inks for use on matt-coated papers should be used, although the only way of reducing the risk of ink rub is to overvarnish (spot varnish) appropriate printed areas, especially heavy solid, tint and halftone areas.

Coated papers are used for a wide range of high-quality commercial printing, especially for work containing halftones, including annual reports, brochures, leaflets, booklets, folders and books, with grammages of from 60 to 200g/m^2.

Chromo paper

This is a high-grade coated paper, usually one-sided and having a thicker coating than art paper. Its main use is in sheet-fed label printing.

Enamel paper

Enamel paper is one-sided and has a highly polished finish. Cheaper grades (box enamels) are used for box covering and better grades for some types of label printing. It is available in a variety of colours.

Cast-coated paper

Cast-coated paper has a uniformly very smooth surface with a high degree of gloss. Cast-coated applies to the manner of obtaining the smoothness and gloss when coating one side of the sheet at a time. The coating method is of a conventional form, usually airknife on a blade pre-coated base paper. A special formulation is used for giving a high coating weight with special properties for obtaining the finish. The wet-coated web is brought into contact with a large heated drum. While wrapped around it, the coating is dried and the surface is a cast of the highly-polished chromium-plated, or stainless steel, drying cylinder. A sheet may be cast-coated on one or both sides. Grammage range is from 70 to 135g/m^2.

Writings

Writings are produced with a surface suitable for writing with pen and ink, but this term is a very comprehensive one because there are many different kinds of paper used for this purpose. Generally speaking, writing papers are more carefully made and, therefore, more expensive than printings. More sizing is required to take writing inks and some better quality papers are *tub-sized* (ts) after manufacture, while cheaper qualities are *engine-sized* (es) in the stock preparation. Extra sizing adds firmness to the sheet and gives additional strength as well as providing greater resistance to ink penetration.

Writing papers may be tinted or white (designated from cream to high white) and may be laid or wove, depending on whether the paper is watermarked with laid lines or not. As with other papers, quality and colour depend on furnish but, in addition, writings may be graded according to the kind of sizing treatment.

Banks and bonds

Banks and bonds were originally produced for typewriting, for which a matt surface is desirable, but can be regarded as writings, because they are used for letterheadings and other office stationery, being suitable for both typewriting and handwriting with pencil or ink. This kind of paper, containing more sizing, is crisper, tougher, and more translucent than a writing paper of the same grade and is also produced in lighter substances than writings.

Banks and bonds are distinguished simply by grammage. If they are of a grammage of $63g/m^2$ or more they are classified as bonds, whereas if they are less than $63g/m^2$ they are classified as banks. Better grades are still made partly from rags and are tub-sized and watermarked, but wood is more commonly used for most grades, the cheaper grades being only engine-sized. They are generally supplied in a variety of colours as well as white, which is often referred to as *cream wove*. Copier papers are similar to bonds, but are supplied to a closely-controlled moisture content.

Manifold is a very thin bank paper, usually about the substance of $30g/m^2$ and sometimes called *flimsy*. It was in popular use when many carbon copies of typewritten or handwritten matter were required to be made at the same time. *Airmail* is a lightweight paper of better furnish, but similar weight and its name indicates its use.

Speciality papers

These are papers intended for specific purposes. Examples include:

Carbonless/self copy paper, where a range of white and coloured stock has been specially coated to produce an image in blue or black when pressure is applied. It is made up into sets with a top sheet coated on its underside, middle sheets coated differently on both sides, with the bottom sheet coated on the face only. Papers are available in sheet and reel form in grammages ranging from 40 to $240g/m^2$.

Gummed papers are available as dry gummed, which has to be wet for application, and are available in a wide range of finishes including mf, cast-coated, chromo and coloured. A wide variety of self-adhesive papers and foils is also available in sheet and reel form, suitable for offset litho, letterpress and screen printing. Surface finishes include uncoated, opaque, machine-coated, matt-coated, gloss-coated, cast-coated. Self-adhesive vinyls are also available in sheets and reels suitable mainly to screen printing in matt and gloss finish, white and colours.

Other speciality substrates include fluorescent, cloth-lined, mottled parchment, metallic/foil surface, glassine/vegetable parchment and embossed finishes.

Different types of boards

The previous notes have referred predominantly to paper, although a considerable number of different types are also available in board weights, which tend to start at $200g/m^2$, weights below this being classified as paper substance.

Pulp board is manufactured in a single web-like paper and has an underside and topside. Twin-wire pulp boards are even-sided, being formed from two webs on the machine. Quality varies with furnish, which may be mechanical, woodfree, or a mixture, and the finish is usually matt or supercalendered. White and tinted boards are produced from approximately 200 to 750 microns.

Paste board is more rigid than pulp board, having a middle of the required thickness lined on both sides with white or tinted lining paper. It is produced in thicker substances than pulp board.

Triplex board is made up of three layers while *duplex board* consists of two plies or webs which are similarly combined in a moist state on the machine, but differ in quality or colour.

A paste board should be distinguished from, say, a white-lined folding box board made on a cylinder-mould machine in which one vat contains a white pulp while the others may have recycled or mechanical pulp, all of which come together to form one web on the machine. Caliper ranges from 280 microns up to as high as 2500 microns.

Coated art boards are pulp or paste board coated on one or both sides.

Cloth-lined and *cloth-centre board* are used when extra strength is required. The former consists of a board backed with linen canvas or linen and the latter of three layers with the cloth at the centre lined on both sides with thin board or paper.

Strawboard was traditionally made from straw pulp and is a solid inexpensive board used mainly in bookbinding, although it is used by some general printers as backing boards for glued and padded sets.

Millboard, greyboard and *chipboard* have similar uses but are superior in quality to strawboard.

Paper/board specifications

Paper suppliers now provide a wide range of technical data on paper and board in price lists and promotional literature. As previously indicated, the performance of a paper or board is governed by the sum of its properties related to a specific purpose or use.

The printer has to consider the *runnability* of the paper or board to ensure as high a level of productivity as possible, as well as print quality, ensuring the required quality result is achieved.

The main properties which affect print quality include: grammage, thickness, bulk, smoothness, brightness, gloss, opacity, sheet formation, dimensional stability, tensile strength and moisture content. For further information and a greater explanation of this area, including the relevant standards, appropriate testing equipment and units of measurement, consult the BPIF publication *Paper specifications and the paper buyer*.

Printing processes and paper/board

Each of the different printing processes require certain characteristics in a substrate to function effectively and to meet the price and/or quality standard required.

Sheet-fed offset litho requires an uncoated paper/board which is well engined- and surface-sized with a firm surface and little loose fibre. Coated stocks do not normally have any problems with loose surface fibres. Long-grain paper gives more dimensional stability.

Heat-set web offset has an upper limit of paper grammage of 135g/m^2 when it is to be folded, with little limitation on sheeted work. Coated paper ideally should have a low moisture content as blistering of the paper surface may occur as the paper passes through the drying oven. *Cold-set web offset* requires a soft-sized absorbent paper.

Gravure is well suited to the cheaper grades of paper such as sc mechanicals, as long as they have a satisfactory smoothness. Web ribbon folding is often restricted to no more than 90g/m^2.

Screen printing can print on virtually any substrate including paper, board, plastic, glass, metal and fabric.

Letterpress requires a compressible, smooth, gloss-coated surface to print fine halftones.

Flexography prints equally well on coated, uncoated or plastic film.

Laser printing paper requires to be curl-free with a low moisture content, long grain and a substance of 60g/m^2 plus.

Suitability of various substrates for a wide range of printed work

Job	Substrate recommended	Printing process	Advantages	Limiting factors
High-quality illustrated work	Art (off machine coated)	Litho and gravure	Opaque: smooth glossy surface; screens up to 60 and above (litho) best qualities; sharpness of definition; used for halftones in monochrome and colour	Relatively expensive and heavy; is liable to crack when folded, especially heavier weights of 135g/m^2 and above
	Matt art, coated cartridge matt, satin, silk		Opaque: eggshell finish; preferred where effect is more important than definition	As above; requires special care in printing; screens up to 80 and above (litho)
Lower-quality illustrated work	Machine-coated printing	Litho, letterpress and gravure	Opaque, can take up to 54 screen and above (litho) according to quality of body paper, folds better than brush-coated art; cheaper than art	The less mechanical to chemical wood used in body paper, the better quality and appearance, but the higher the cost
Quality work	Cast-coated	Litho, letterpress and gravure	Opaque; can take fine screens	Expensive
Inexpensive illustrated work	Machine-finished (mf) and sc	Litho, letterpress and gravure	Machine-finished takes up to 34 screen and above (litho); good grades supercalendered will take up to 40 screen and above (litho)	Inferior mf grades show wire marks on underside, which makes good printing of 40 screen if well printed halftones difficult
Typematter without half tones	Antique wove	Litho and letterpress	Gives high bulk	Generally unsuitable to halftone illustrations
Inferior work with or without illustrations	Newsprint	Litho, letterpress and flexo	Low price; will take screens 26 to 34 and above (litho) according to finish	Not durable; cheapest grades discolour rapidly
Stationery and forms	Parchment bond, bank, wove, laid and airmail	Litho and and letterpress	Accept writing ink well Saves weight and bulk for overseas correspondence	Printing is a little more difficult compared to other papers
Envelopes	As above, plus cartridge, manilla, etc	Litho, flexo and letterpress	Strong and relatively inexpensive	Less expensive grades not so strong
Books, account books and loose-leaf	Wove, cartridge, laid	Litho, letterpress and flexo	Strength and opacity	Generally unsuited to tints and halftones
Greetings cards, postcards	Pulp, coated boards	Litho	Strength, rigidity and bulk, varnish or coating applied for higher finish	Generally restricted to one-sided material
Cartons	Folding box board, pulp, coated boards	Litho, gravure, flexo and letterpress	Strength, rigidity, stiffness, bulk coating or varnish applied for higher finish	'Specialist' carton material required
Self-adhesive	Coated, uncoated	Flexo, screen, letterpress	Wide range of sheet and reel material with good quality printing properties	Expensive: need to ensure correct adhesive properties
Flexible packaging	Plastic film Specialist papers	Flexo, screen	Wide range of specialist materials with good quality printing properties	Specialist materials only suitable

13 Sourcing suppliers and purchasing

What products and services to provide, using in-house resources, or alternatively outside suppliers, is of constant consideration and deliberation to printing companies in trying to satisfy a wide range of customer needs and requirements.

In the production of printed matter, printers process a wide range of 'raw' materials, semi-processed and fully-finished items. 'Raw' or unprocessed materials purchased will include paper, board, ink, photographic paper and film, processing chemicals, planning foils, plates, cylinders, stencils, adhesives, stitching wires, thread, leather, rexine, cloths, metallic foil etc. Semi-processed and fully-finished items are also extensively used in the printing industry and can be purchased through trade houses as and when required.

The areas covered in this chapter are as follows: supplier selection/approved suppliers - past performance, quality assurance; supplier appraisal - inspection and testing, quality audits, supplier rating lists; materials requirement planning; JIT (*just-in-time*); EDI (*electronic data exchange*); partnership sourcing; outwork - the need for trade houses, services and products provided by trade houses, the account executive's responsibilities, pre-planning and placing orders for outwork.

The overall material costs incurred in producing any printed matter can be as high as 50% or more, it is therefore important that an account executive is fully aware of their company's policy re sourcing suppliers and purchasing and understands how and why they choose certain suppliers over others.

Figure 4.4, on page 56, takes the form of a flow chart illustrating outwork purchasing and processing procedures.

The purchasing situation is constantly being reviewed within companies and it is an area in which account executives can make a positive contribution, by ensuring sound and positive feedback between both parties through their experience of handling and processing work on a day-to-day basis.

A printing company needs to set up a set of criteria which will result in the right materials, goods and services being available at the right time, from the right supplier at the right price and in the right place.

Purchasing materials, products or services at the cheapest price possible is seldom the sole criterion, as although price is of major concern in any purchasing consideration, quality in the form of 'fitness for purpose', availability and consistent high standards of service are also of paramount importance.

Supplier selection/approved suppliers

It takes a great deal of technical and commercial awareness, time and effort to build up a nucleus of 'approved' suppliers which can regularly and consistently meet the needs of printing companies, all of whom have similar, although not necessarily identical requirements.

The current approach adopted by most companies is to set up a *supplier appraisal procedure* which results in the successful applicant being approved as a supplier.

Before being accepted as a recognised or approved supplier to a printing company, each vendor needs to prove it possesses the ability to consistently supply goods and/or materials which meet the necessary quality, commercial and technical standards required.

Figure 13.1 illustrates an example of a *supplier questionnaire form* which a printing company would send to a potential new supplier such as a paper merchant or a trade house; alternatively, the printer could, in fact, be a recipient of this form from a prospective new print buying customer.

Past performance

In drawing up a list of approved suppliers, printing companies will often base their assessment at least partly on *past performance* - ie how well a supplier has performed in the past in meeting its needs.

Quality assurance

Another major method of assessing whether a supplier is worthy of consideration for acceptance as an approved supplier is in its approach and commitment to quality assurance, as demonstrated by the company operating under a process-controlled environment.

Quality assurance is defined as the process embracing all activities and functions concerned with the attainment of quality.

A *quality system* is defined as the organisational structure, responsibilities, procedures, processes and resources for implementing quality management. A supplier's or printer's quality system therefore provides the framework which sets up, monitors and controls quality. If the quality system is formalised, such as required by BS5750/ISO9000 and approved by a recognised certification body for this purpose, then it gives the ordering company confidence that adequate procedures are in place within the supplying company's organisation to assure a satisfactory transaction as far as possible.

SUPPLIER QUESTIONNAIRE

Company name: ____________________

Address: ____________________

____

Telephone no. ____________ Fax no. ____________

Quality representative of company: ____________________

Person responsible for completing
this questionnaire: ____________________

Quality system approval

Standard approved to	BS5750 part 1	part 2	part 3
	ISO9001	ISO9002	ISO9003

Registration no. ____________ Scope of approval ____________

Accreditation body ____________ Date of approval ____________

Do you have a quality manual?	Yes/No
If requested, would a copy be made available?	Yes/No
Is the quality system audited?	Yes/No
Do you verify incoming materials?	Yes/No
Do you maintain records of in-process and finished product inspection?	Yes/No
Do you have a system to identify, segregate and control non-conforming items?	Yes/No
Do you operate a corrective action system?	Yes/No
Can you supply a certificate of conformity on demand?	Yes/No
Do you regularly calibrate inspection and test equipment used by your company?	Yes/No
Are calibration records maintained?	Yes/No
If requested, would you allow access to your quality system by our representative?	Yes/No

Figure 13.1: Supplier questionnaire form

BS5750 Part 2/ISO9002 sets out requirements where a company is manufacturing goods or offering a service to a published specification or to the customer's specification - *this is the Standard to which most printing companies apply for certification.*

Under BS5750 Part 2 the following requirements are set out against purchasing: *4.5 Purchasing,* sub-divided into 4.5.1 *General,* 4.5.2 *Assessment of sub-contractors,* 4.5.3 *Purchasing data* and 4.5.4 *Verification of purchased products.*

This section covers the requirement to ensure conformity to stated specifications and that the selection of suppliers (sub-contractors) should be based upon previously demonstrated capability and performance. In addition, clear and unambiguous purchasing documents need to be used ensuring precise identification of product, also the purchaser has the right to review and approve that the purchased product conforms to agreed specifications.

Supplier appraisal

Supplier appraisal, also known as *vendor rating* is the assessment, approval and review procedures undertaken in order to check the performance of suppliers. To ensure suppliers are providing a consistent, satisfactory, quality service, it is necessary to formalise the process and to record an assessment of their performance against certain criteria over a period of time, with a review at certain intervals, say every six or 12 months.

The assessment may take the place of comparing and contrasting a range of vendors providing a similar service against the checks required to ensure the work is up to standard.

Inspection and testing

This is often undertaken on the basis of a checklist procedure at all stages of production such as initial receipt of material, intermediate stages and final printed copies.

Examples of such tests would include the supply of substrates, such as paper, board or vinyls, etc, presensitised offset litho plates or colour repros being checked against the required specification for each item.

Checks would be undertaken on receipt of each item for condition of packaging and goods, on delivery note details such as quality, type, size, specification, purchase order number and instructions, plus checks throughout the printing process up to and including final inspection, approval and despatch of printed matter to the customer.

Quality audits

These are undertaken at regular intervals to check that the quality system is operating satisfactorily, including identifying problems associated with suppliers.

Supplier rating lists

An additional tool of supplier appraisal is the use of supplier rating lists. In this method, for the system to work effectively, it is essential that printing companies make it clear to their suppliers what they require in terms of product, service, quality, specification standards and tolerances, working practices, inspection and testing, payment terms and purchasing procedures, etc - ie raising any points which the purchaser considers is necessary to ensure work is carried out exactly as required.

As part of a company's purchasing procedures it is common for printers to produce a simple table such as *Figure 13.2* which is used to record a supplier's performance against set criteria. The supplier is informed of their performance rating at least once a year; if there is a problem, however, this will be dealt with immediately and could well result in a supplier being withdrawn from the approved supplier category. In return, suppliers are often invited to comment on their rating and to communicate on how the situation could be improved on both sides.

SUPPLIER APPRAISAL AND REVIEW

Criteria	Maximum	Awarded
Overall consistent quality	20	
Adherence to specification standards and tolerances	20	
Price competiveness	15	
Meet delivery and target times	15	
Speed and efficiency in rectifying faults	10	
Communication and feedback	10	
Supporting documentation	10	
	100%	

Figure 13.2: Supplier appraisal and review form

MRP (materials requirement planning)

Stockholding of any kind is expensive in terms of the space taken up and capital tied up in stock, but printing, like any other industry, requires to maintain at least a minimum stock level of raw materials, apart from partly finished items in the form of work-in-progress.

Paper/board and ink are the main items held by printers. Others include bromide paper, film, plates, chemicals, adhesives, wire and cloth. The material content of a printed job can account for 25 to 50% or more of the total cost.

It is essential, therefore, that the size of all inventories is kept to an absolute minimum commensurate with the usage level/availability. Selecting the variety of materials to be held in stock - particularly paper, board and ink - should be the combined task of senior representatives, production management and materials buying department. Most suppliers today have regional warehouses stocking most printers' needs and can deliver the majority of paper, boards and inks within 24 hours of order, sometimes even earlier. Other materials may take up to a week from placing an order, but *forward planning/ usage forecasting* can make an important contribution towards a healthy and profitable printing company.

Two different types of materials ordering can be identified -

i) *materials ordered for a specific job or number of jobs* for which the requirements are known.

ii) *materials purchased in anticipation of orders* which have not yet been obtained but where experience indicates a continuing or regular future requirement - eg stock paper for regular magazines, presensitised positive and negative plates. Materials not held in stock but needed for a specific job will be ordered when the works instruction/order set is raised, in the quantity required for the job, and this should be part of a simple routine.

A measure of the effectiveness of the use of stock is the number of times it is sold and 'turned over' in a year. This is termed the *stock turnover* - eg if the average value of stock held is £100 000 and the cost of the materials used in the year is £550 000, then the stock turnover would be 5.5. The average printer appears to turn over stock about six times a year. The aim must be to maintain this figure at its highest possible level without reducing the service to the customer.

Chapter 2, The administration and processing of printed matter, pages 30 to 35 covers the subject of stock control procedures.

JIT (just-in-time)

JIT (*just-in-time*) is a strategy which aims at ensuring raw materials and work-in-progress are kept to a minimum by efficient forward planning and forecasting. At its simplest, just-in-time is aimed at ensuring the right materials and quantities are produced and delivered to the required production points/destinations, at a designated time.

When operated effectively, JIT involves everyone in the production chain from the final user - ie the customer requiring printed matter, through the various printing production stages requiring raw materials such as paper, board and ink plus outwork items, including for example colour repro, gravure cylinders, cutting-and-creasing dies.

Taken to perfection, JIT results in zero inventory with no waste of time or materials. To operate with such effectiveness there needs to be a continuous flow of carefully-managed, meticulously-planned, timed communications and instructions between all contributors - both internal and external to the company - within the production chain.

It is essential that a printing company can control its own internal production system within clearly defined control standards, before it becomes part of a wider controlled production chain.

The printing process covers a wide range of operations which are often difficult to predict to very fine tolerances, but as printers get closer to their suppliers and customers then JIT strategy and implementation becomes a more practical and realistic proposition.

EDI (electronic data exchange)

A further development in closer supplier-customer relations which includes the sharing and accessing of data for control and information purposes is the use of EDI *(electronic data interchange)* which is the electronic transfer of data between computer systems in different companies. Such facilities are bound to grow as suppliers and customers seek faster and more accurate ways of transmitting and receiving information from each other.

Two distinct aspects of EDI can be observed: firstly, it allows products to be described, specified, acknowledged and invoiced electronically. Printers can thus order raw materials using EDI, while their customers might describe a print job by using an 'electronic ordering or estimating form'. Secondly, the computer link would allow one party to interrogate another's cost rates, stock holding and price lists. Obviously, in this case, a decision has to be taken on what information can be accessed by outside parties.

Partnership sourcing

The definition of partnership sourcing is that it is a commitment by both customers and suppliers, regardless of size, to a long-term relationship based on clear, mutually-agreed objectives.

The key objectives of partnership sourcing as set out by Partnership Sourcing Limited operating through the CBI are:
- to minimise total costs
- to maximise product and service development
- to obtain competitive advantage.

Partnership sourcing is based on the concept that joint ventures carefully identified and researched in terms of strategy, overall commercial advantage and improvement, and expressed as common goals agreed between the parties concerned, leads to mutually beneficial business relationships.

The benefits of this approach to printing companies can be considerable as, generally, print procurement by customers is very much based on short-term agreements, if any agreements exist at all, other than for say contracts renewed annually for magazines and publications, or the supply of a certain quantity of cartons or labels over a period of time depending on demand.

Improved working relationships, along with a more focused and targeted strategy, should lead to reduced purchasing administration time and costs, plus more efficient production planning. In turn, this should lead to an improved understanding of customer requirements where it can be recognised in a positive way by some form of supply partnerships or sourcing.

Outwork

'Trade houses' is a term often used to describe the vast number of suppliers and specialist services used by printing companies. The suppliers of materials for stock, or of equipment and machinery, are not included in this description.

Account executives place work with trade houses when the required skills or production processes involved are outside the capacity of the organisation's own resources, or when there is an economic or production advantage such as a bottleneck in production departments which cannot be resolved using only in-house facilities. Account executives will also often be called upon to order paper, board or other raw materials required for jobs.

Services and products provided by trade houses

Trade houses cover all kinds of pre-press, printing and finishing processes. The extent and variety of the services available is considerable, as the following list shows:

Adhesive binding
Artwork studio
Bar coding
Blister packaging
Blockmaking
Bookbinding
Boxmaking
Cylinder engraving
Colour printing
Colour proofing
Colour repro
Colour scanning
Continuous stationery
Cutting and creasing
Cutting and creasing dies
Diecutting
Diestamping
Direct marketing services
Disk conversion
DTP bureau
Duplicate book manufacturing
Encapsulating
Engraving
Envelope manufacturing
Eyeletting and stringing
Fancy stationery manufacturing
Folding and slitting
Gold blocking
Gravure cylinders
Gumming
Imagesetting bureau
Index cutting
Jacketing (dust jackets)
Laminating
Laser engraving
Laser generated dies
Loose-leaf book manufacturing
Mailing
Map printing
Metal printing
Offset platemaking
Paper/board graining and embossing
Plastic-comb binding
Plastic welding
Planning - conventional
Planning - digital
Photography - conventional
Photography - digital
Punching and perforating
Round cornering
Ruling
Scanning - mono and colour
Screen printing
Showcard mounting
Spiral wire and plastic comb binding
Step-and-repeat
Thread sewing
Ticket and tag manufacturing
Trade printing
Typesetting
Varnishing
Window aperturing
Wrappering (direct mail)

The need for trade houses

The existence of, and the necessity for, trade houses is due primarily to the structure of the printing industry itself. As explained in *Chapter 1, The printing industry today*, the average printing company is very small in terms of the number of employees with over 75% of BPIF member firms employing less than 25 people at September 1993.

On this basis, therefore, many printing companies cannot afford to install expensive equipment for special processes which will not be fully occupied. Even if a company is prepared to invest capital in machinery to carry out certain finishing processes, there is such a wide variation in the requirements of customers that it would be impossible to install equipment capable of satisfying all possible demands.

The trade house, on the other hand, because of much greater volume of demand, can install machinery to cover a great variety of requirements within its own specialist field. Individual printing companies, when faced with orders requiring any kind of special skill or particular machinery outside their normal run of work, can offer their customers a comprehensive service by calling upon these specialist trade houses to fulfil such parts of the work as may be necessary.

The account executive's responsibilities

The first responsibility of account executives is an awareness that it will be necessary to place certain elements of the work with a trade house. An example of this situation would be where a printer produces small quantities of adhesive bound work in-house on semi-automatic equipment, but for long runs uses a trade finisher with fully-automatic facilities.

The choice of a suitable supplier of the services required is obviously an important consideration. This will often be conditioned by the closeness of the printing works and the trade houses. Obviously, the nearer the better. Consideration must also be given to the relevant production schedules, including not only the time required for production of the job, but also for transporting the material to be processed to and from the trade houses. The means of transport and the costs involved also need to be taken into account.

Complete reliance on a single supplier for specific types of work is not necessarily the best policy, since by distributing enquiries or orders, and particularly large ones, the opportunity occurs to compare prices, quality and service.

On the other hand, having established a sound satisfactory relationship, regular orders placed with the same range of companies enhance that most desirable element - *service* - and it may sometimes be possible, in repeat orders, to dispense with the need for obtaining quotations in advance.

Reference should be made to the possibility that printing groups with two or more factories may have facilities in one which do not exist in another. It would clearly be in the individual company and group's interests to keep the production within its own organisation, wherever possible and practical, each company using the others in much the same way as they would use trade houses.

Pre-planning

As account executives will be dealing with methods and processes which are often not available in their own works, it is essential that preliminary discussions take place with the trade houses to discuss what is to be done within the company's resources and to co-ordinate this for the process or machinery to be employed by the outside supplier.

For example, if a cover with glued-in pocket and gussets, is to be printed sheet-fed offset litho four-up and to be cut and creased by an outside trade supplier, what size and type of machine will be used in both processes? What will be the laydown? What allowances will be required for gripper edge, bleed, gussets, distance between cutting blades and waste stripping areas?

Such considerations are essential before proceeding to platemaking and, of course, could apply equally to trade UV spot-varnishing on previously printed covers. Further points could be - can the job be printed eight-up or is that too large for further processing? If so, should the printed sheets be cut first by the printer if they are too large for the trade house to handle, or is there a risk that a problem of maintaining accurate register will be involved if the gripper or lay edge is lost by cutting the sheets in half?

When the job in question is a bound publication requiring trade binding, then there is the question of the appropriate imposition scheme for the trade bookbinder's folding machines; or possibly how the printed design for the cover of a book or the image for printing the endpapers should be laid down in the sheet. The question of the direction of the grain of the paper is also very important.

Placing orders for outwork

All orders placed for outwork services should be on official purchase order forms, numbered, usually in triplicate, with each order being assigned a reference number, which should be used in all correspondence. The top copy is sent to the supplier, the second to the costing department and the third retained by the originator of the order - eg the account executive.

With the introduction of Management Information Systems, as raised in *Chapter 3*, more and more printing organisations are using their system to generate their purchase order documentation.

In all purchasing procedures it is important to ensure that the specifications are absolutely clear. Nothing should be left to chance or guesswork. Where orders are written by hand they must be legible and, where estimates have been submitted, these should always be referred to. Despatch instructions are also a most important part of the order.

In circumstances where the delivery is to be made direct to the customer from the trade house, instructions regarding packing and labelling should not be overlooked. In these circumstances it may be necessary for the account executive to send the appropriate labels and delivery notes as normally used by his/her own organisation.

It cannot be assumed that because an order has been placed and the material or partly finished goods delivered that the job can be forgotten. 'Out of sight, out of mind' for some, but never for an account executive - the expression 'if you can't stand the heat, stay out of the kitchen' seems particularly apt when considering an account executive's job!

Where the supplier has, for example, to make special dies or cutters it is important to send a tracing, approved proof or printed sheet as soon as one is available, so that the preparatory work can be carried out in advance.

Finally, it is a wise precaution to arrange for advance copies to be delivered for inspection to the person responsible for ordering the work, before the main delivery is carried out. After delivery is completed, the invoice, quantity, and price must be checked, and forwarded to the accounts office, via the costing clerk, for payment.

It is virtually impossible for anyone to be knowledgeable on all the technical details involved in the processes and finishing methods employed throughout the printing industry. An efficient account executive nevertheless makes it part of his/her job to know, or discover, where these services exist, to which kinds of job they are applicable or necessary, and to discuss and arrange for the operations to be carried out, as may be required for specific orders.

14 Customer service

Customers are the vital ingredient all companies need - without customers no company can exist. Treated well, with respect and a genuine interest in serving them, customers are more likely to respond positively, building strong business bonds and relationships: however, treated badly, without understanding and empathy, the customer will find alternative suppliers more aware of their needs and requirements.

Printing is a bespoke industry relying on producing printed matter to a customer's specific requirements; printers therefore have very little influence on the demand for printing. It is an extremely competitive industry, becoming even more competitive as all industries slim down and become leaner. To address this situation, printing companies have been forced to become more aware of the level of service they give to customers and to become more knowledgeable about their customers' businesses.

This chapter sets out some of the approaches successful companies have undertaken in recognising the importance of customer service to ensure not only the retention of their existing customer base, but also its expansion through customers becoming business partners and advocates.

The topics covered are customer service feedback, customer service is everyone's responsibility, customer service programme, aims of customer service programme, senior managements' role in customer service; issuing and receiving instructions - preparation of the works instruction ticket.

Customer service, also often referred to as *customer care,* is a strategy which must permeate the whole of the company's philosophy so that it becomes customer-centred.

A considerable proportion of the printing industry operated previously as an 'order taker', where customers would place orders with printing companies without too much effort being expended by the printer on customer support and service activities.

Trading circumstances were such that many printers had a client base where a high proportion of customers had been placing orders with them, for several decades, and therefore felt dependent on them. This was partly because of the support they had received, but also due to a lack of alternative suppliers with which they could compare and contrast the service received.

A further contributory factor was that the printing industry appeared to be a 'black art' to many print buying customers, so they were reluctant to change arrangements which had been in place for some considerable time and appeared to work perfectly adequately.

The vast changes which have taken place in business relationships, trading conditions and technical advances in recent years, have forced all businesses to completely rethink their relationship with customers, because customers are, in essence, the lifeblood of any organisation.

Printing, as an old and established industry, has not found it easy to adapt to the much more aggressive business world of the 80s and 90s. Printers, to survive and prosper, have had to become 'order creators' rather than 'order takers', playing a much more positive role in selling their expertise. This is in contrast to the largely passive role which marked a considerable part of the previous approach by many printing companies.

This book has highlighted the technical changes which have 'opened-up' printing in so many ways, so that customers requiring printed matter have many alternatives on how they can organise and plan their print requirements.

The account executive as highlighted in *Chapter 4, The role of the account executive,* is a focal point of contact for customers and it is important that he or she understands the role customer service plays in an organisation, and their place in it.

Customer service feedback

The essential drive behind customer service is in providing goods and services, to a consistently high standard, which meet the needs and requirements of customers.

It is a strategy which is constantly under review, centred around customer set criteria and reinforced with ongoing feedback from customers on how they rate the printing company's customer service.

Part of the feedback mechanism is often the use of a *customer service questionnaire,* similar to *Figure 13.2* on page 229, only using topics centred around customer satisfaction generated through customer service.

Other examples of increasing customer response rate and contact include the following:

Freephone/helpline - this telephone facility is particularly useful where the printing company is selling a range of stock line products such as office stationery, business forms, labels, leaflets, brochures, etc which will often not be bespoked to individual requirements. Companies offering this service will often operate at least part of their business through a catalogue which allows customers to order by telephone or fax, and this same facility will be used for further information and assistance. The data built up from customers' purchasing habits will provide valuable information which will be used to monitor and control stock lines, also to gauge customer service response from questionnaires and subsequent mail shots.

Customer service teams - these are formed to initiate and co-ordinate customer care projects and programmes. Groups are formed in much the same way as quality circles or teams within a TQM *(total quality management)* environment. Representatives will normally be made up from each department or function, reviewing and discussing how customer service can be improved in their own area, as well as ensuring all efforts are co-ordinated to the overall aim of an increasingly high level of customer satisfaction.

Customer support team visits to clients - frequently the only person to visit customers in situ is the outside sales representatives where most of the discussion will tend to centre around specific enquiries or jobs. It is important therefore to find another vehicle - ie point of contact, where the customer support team can visit the client to discuss the quality of service they are receiving and how this can be improved. It will only be necessary for a small team to visit on a regular basis, say every six months or so, with a qualified person from a production area attending, when necessary, to deal with technical matters which have arisen. This arrangement allows printing and customer staff, including account executives, to meet and discuss mutually related areas, which can only benefit the overall understanding on both sides.

Open days and customer/potential customer visits - these are aimed at putting the host company on show. Open days are often held to coincide with a particular event which the printing company wishes to bring to the attention of customers and/or potential customers - examples would be the installation of a new press, other item of equipment, or moving to new premises.

Other customer visits are organised on an individual basis where a single customer is targeted to visit the factory: in this instance it could coincide with their job being run or passed-on-press, alternatively, it could simply be a promotional exercise undertaken with a potential new customer.

Customer response/complaint cards - this operates in a similar way to the customer service questionnaire where the printer is eliciting customer response, but in this case the information is linked to on-going jobs where specific problems have arisen. It is often run in conjunction with the company's quality system procedures where the cards are issued on receipt of a complaint, rejection of a particular job or batch of printing. The information received from the customer is acted upon by the quality representative to rectify and solve the problem; it is also used for customer service response, with often further follow-ups, once the problem has been resolved.

It is important that customer service responses from all approaches and points of contact are measured and acted upon, in the same way as for other business data and ratios, such as financial and production.

Only by applying and retaining the drive towards customer service excellence, monitoring and reacting to the findings, will the system work effectively.

Customer service is everyone's responsibility

In the same way that quality systems rely on the concept of *internal customers* - ie everyone relying on everyone else in the organisation to ensure they complete their job function in such a way that the work carried out by them follows on to the next stage exactly as specified and instructed - so the same philosophy exists within customer service.

Everyone has a part to play and everyone must play their part - customer-centred attitudes therefore need to permeate the entire staff of an organisation.

It is the collective response of everyone's efforts to understand the customer's needs and to satisfy them, that ensures each printed job represents a further cementing of the relationship between printer and customer.

As previously indicated, customer service and care is the developing and nurturing of personal relationships with every customer, reflecting a company policy which recognises that the most important part of any business is the customer.

If properly implemented, customer service can ensure that satisfied customers are loyal customers who develop into advocates of the company, so building and widening its customer base.

For customer service to be effective, printing companies must spend time gaining an understanding of the markets and business in which their existing and prospective customers operate. By doing this, they will gain a market advantage over their competitors.

Customer service programme

A customer service programme is set up to ensure that customers will find doing business with a particular company a friendly and rewarding activity, and will want to repeat the experience.

It is essential when introducing and running a customer service programme that customers needs and expectations are established; without this information the programme would lack validity and purpose.

Targets/standards of service need to be set and monitored carefully for different functions and departments.

This setting of targets is an essential element in establishing an effective customer service programme; there must be a level of service to which the company must aspire, maintain and then improve on, in order to build improved business relationships and partnerships.

The targets set out below are not intended to suggest a company should accept anything less than 100% service level in all areas, but only to give a general guide at where one might start a customer service/programme - discussion and consultation between all groups of employees will be crucial, as will be the feedback from customers.

Receptionist - responding to telephone calls within six rings, 100% of the time; responding in a courteous and efficient manner and greeting the caller by name. When the call is put through it is essential that the person requested is given the name of the caller and the reason for the call if this can be established. If the person is unavailable and no other person can handle the enquiry on their behalf, the receptionist should take the full particulars surrounding the enquiry and ensure that the customer is kept fully informed if the person is not available as agreed for any length of time.

Outside sales representatives - attending customer appointments 100% on time, and a success rate of 25% conversion rate from quotations submitted to orders placed.

Estimators - 50% of estimates received completed in one day, 95% within two days.

Account executives - 50% of works order documentation completed in one day, 95% in two days from confirmation of order.

Production control - 90% of work delivered on time.

Quality department - reduction of customer complaints by 10% every six months.

Production departments - reduction in spoilage and reworkings by 10% every six months.

Van driver - 100% of printed matter delivered when and where the customer wants it, along with appropriate documentation duly signed off.

It is important to recognise that the service given by receptionists, sales representatives and van drivers has a considerable influence on how a customer perceives a company as they are a regular, often face-to-face, point of contact with customers.

Procedures must also be set up and followed to ensure the customer service programme is fully integrated into the company's operations.

Figures 4.2, 4.3 and *4.4* cover the stages from *enquiry/order* to *order admin/job entry; order admin/job entry* to *order documentation to production/ service departments* and *order outwork purchasing*. It is by adopting procedures such as these and others which cover the whole operation, that a company is able to control the printing process into predictable stages providing information and feedback on achieving the targets set on customer service.

Aims of customer service programme

- to improve retention of existing customer base, expanding the size of each account where possible

- to introduce new customers through recommendation and reputation

- to gain a market advantage over competitors

- to improve customer loyalty and company image, both to employees and customers

- to improve value added (see pages 59 and 60) return on jobs

- to increase the level of customer service on an ongoing basis

- to make the business more 'user friendly' and successful.

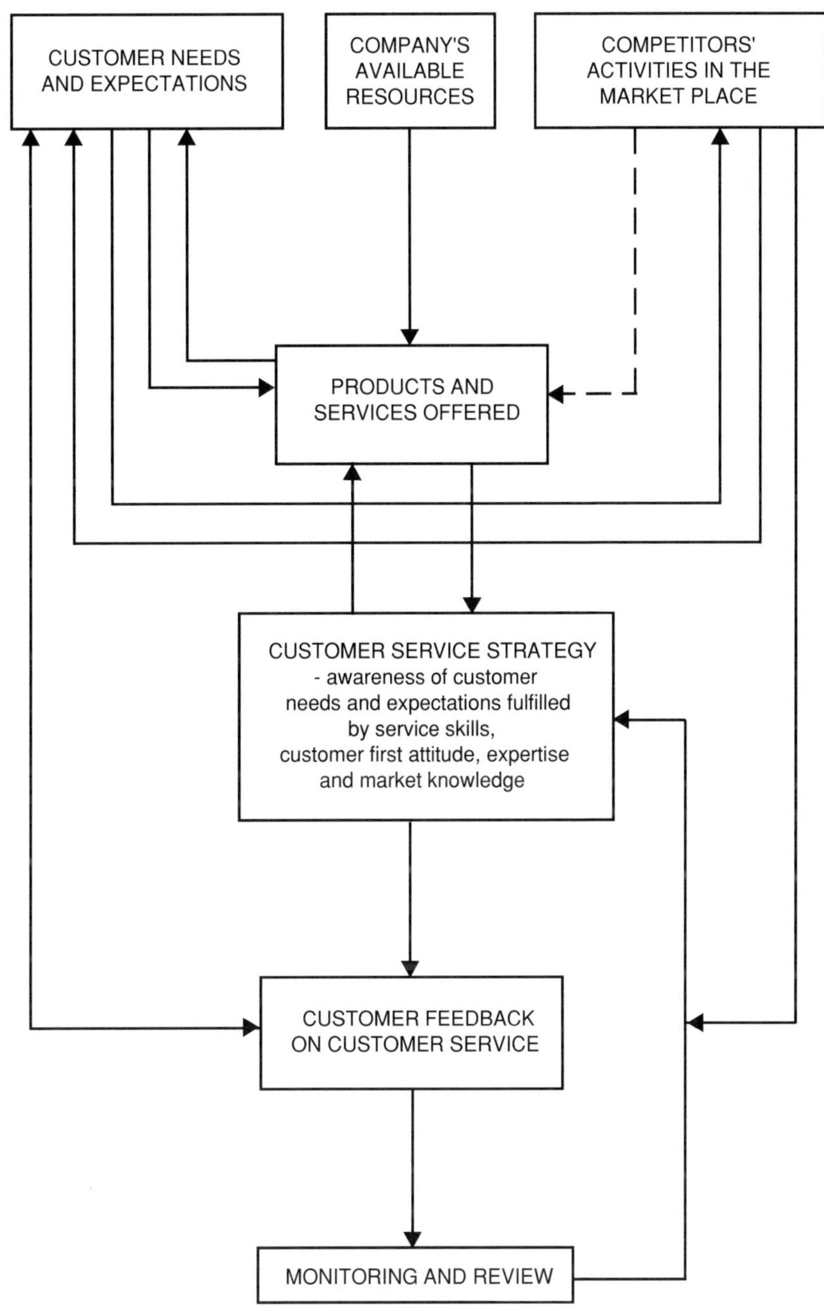

Figure 14.1: Flow-chart illustrating customer service programme

Senior managements' role in customer service

Senior management need to communicate clearly their vision or mission of what the company stands for in a language and form which everyone can understand. They are responsible for contributing to, and establishing key areas as follows:

The reasons for adopting a customer service programme and the benefits to be derived from its installation.

Commitment required by everyone to implement the programme successfully.

Personal and departmental targets set and reviewed at regular intervals.

Training, coaching and support programmes set up to inform and develop knowledge, skills and attitudes conducive to good customer service practice.

Establishing a common goal within which everyone can feel part-ownership.

The customer service programme has to be driven from the top down - senior management must continually emphasise the importance of a customer first policy.

Issuing and receiving instructions

An important aspect of customer service is in ensuring that all instructions received, related to enquiries and job details, is communicated accurately and concisely.

In handling and processing printed matter, mistakes unfortunately do happen such as printing the wrong PMS ink colour or using the wrong paper or board for a job. Issuing or receiving instructions is fraught with difficulty unless it is carried out in a structured and formalised manner.

Customer service as it affects account executives relates to all stages of printed work - ie pre-order stage, order processing stage and after-sales stage.

The following points are included to assist the account executive and others involved in processing print when giving or receiving written instructions.

Written or verbal communications - think first, get your own mind quite clear as to what is required, using terms the other person can understand.

In discussion - listen attentively, keeping an open mind to suggestions, or other points of view, when these may be expressed.

In giving verbal instructions - whenever possible, check the understanding of the one to whom the instructions have been given. Ask if you have made yourself clear, getting the person to repeat anything which may be misunderstood, as well as checking your own understanding of the situation.

Avoiding misunderstanding - it can be part of the account executive's lot, having passed the copy for the job together with complete instructions, to discover later that the job is in serious trouble somewhere in the works because the instructions given were not clear. The anticipation of possible misunderstanding is the key to the whole problem. The job put in hand properly at the beginning stands the best chance of going through smoothly and speedily, to the customer's satisfaction.

The following simple rules, if applied regularly, will go a long way towards the avoidance of duplication of effort, unnecessary problems, and frustrating delays:

1) *Be certain you know what the customer wants.* The possible irritation on account of your persistence will be nothing compared with what the customer will feel later if the job is produced wrongly.

2) *Make sure the production departments know what the customer wants.* They should know everything that you know about the job and this should be in writing. Works instruction tickets are intended for this very purpose.

3) *Make a point of the customer's special requirements,* particularly where these may be different from the normal procedure carried out in your organisation.

4) *Be ready to answer any questions* the production departments may want to ask you without making assumptions - find out the right answer, don't guess.

5) *Report changes or additional instructions* in writing immediately you know about them.

6) *Give yourself time to do the job properly.* If you treat the preparation of instructions carelessly, operatives will tend to be careless too. If you take time to read what you have written and insist on clarity yourself, they will also take care in the production of the job.

Preparation of the works instruction ticket

Having considered the need for getting one's own mind quite clear as to what is required before giving instructions, and the need for communicating adequate information accurately, it is necessary to consider what actions should be taken by an account executive as a preliminary to the preparation of a works instruction ticket - the background to the operation of the works instruction ticket is covered in *Chapter 2, pages 25 to 28*.

The following is a representative list:

1) Send an acknowledgement of the receipt of the order.

2) Check whether the job in question is suitable for your printing plant, or whether additional services or processes (outwork) will be required.

3) Check whether the job has already been the subject of an estimate, or is a reprint.

4) Ensure that all necessary copy, artwork, photographs and other material supplied by the customer is both satisfactory and complete.

5) Check whether disks, artwork, film, paper or other materials are to be supplied by the customer.

6) Check that the necessary materials are available, or can be made available in time, or make suggestions as to suitable alternatives.

7) Ascertain details of any special customer's requirements, for example, the number of proofs, packing and delivery arrangements.

8) Determine delivery dates for proofs and finished copies, and ensure that the requirements are possible in the time allocated.

Glossary

This glossary is intended to be used as a means of reference and explanation to expand on some of the technical and operational terms used in printing and other industries generally.

A sizes Main series of finished printing trimmed sizes in the ISO international paper size range.

Access To recall data from a computer storage area.

Achromatic printing Method of colour printing in which any hue is created from two colours plus black, rather than three. An extension of under colour removal (*UCR*).

Acoustic coupler A method of transmitting data over telephone lines or radio links using the microphone and ear piece to transmit and receive an audible digital sound. To transmit computer data over telephone lines, the data must be converted into electrical tone signals, and sent at a comparatively low rate (see *modem*).

Additive primaries Coloured lights in red, green and blue, (R, G, B) which when combined with each other in equal proportions produce white. Other colours may be produced by mixing different proportions of each light source. Video monitors use this principle to produce colour television images. Input scanner detectors sense red, green and blue components of the scanned image before electronic conversion to printing colours of cyan, magenta, yellow and black. Output transparency recorders generate red, green and blue from input information generally supplied in cyan, magenta, yellow and black electronically re-coded to red, green and blue.

Adhesive binding Style of threadless binding in which the leaves of a book are held together at the binding edge by glue or synthetic adhesive and a suitable lining.

Air-brush An instrument having a small reservoir to contain liquid ink and so arranged that a controlled current of air is blown over the ink surface which is broken down into an atomised spray and ejected through a nozzle. Used by artists to obtain graduated effects on drawings, photographs and lithographic printing surfaces.

Air-dried Paper dried by a current of warm air after tub-sizing.

Alignment Horizontal positioning of type to ensure that the base of each character is perfectly in line with the next.

Antique (finish) A rough, uncalendered finish applied to paper used for book printing, when bulk and light weight are required.

Art Substrate which has received a coating to the base material. It has a very smooth surface, which may be gloss, matt or dull.

Artwork Text, graphic and illustrations arranged individually or in any combination for subsequent printing. Artwork may conventionally be drawn in black and white on suitable artpaper or board; or may be computer-originated, in which case it may be supplied as digitised data on a floppy disk or other means of electronic data. Artwork may also be in the form of a full-colour drawing or picture which requires specialist reprographic colour separation. This enables the separation to be printed in the four basic printing process colours (cyan, magenta, yellow and black).

Ascender Top part of the lower-case letter stretching above the x-height of the character, as in d, h and k.

ASCII **A**merican **S**tandard **C**ode for **I**nformation **I**nterchange. This is a standard coding system used within the computer industry to convert keyboard input into digital information. It covers all of the printable characters in normal use and control characters such as carriage return and line feed. The full table contains 127 elements. Variations and extensions of the basic code are to be found in special applications.

ASPIC **A**uthor's **S**ymbolic **P**re-press **I**nterfacing **C**odes. Standard mark-up or code system which indicates a change of typesetting format. See *SGML*.

Author's corrections Corrections made by the author on proofs, that alter the original copy. The cost of making such alterations is charged for, in contrast to printer's errors or house corrections.

B sizes ISO International sizes intended primarily for posters, wall charts and similar items where the difference in size of the larger sheets in the A series represents too large a gap.

Back The back of a book is the binding edge. To back a book is to shape the back of a previously rounded book, so as to make a shoulder on either side against which the front and back covers fit closely.

Backing or release paper The component of a pressure-sensitive stock which functions as a carrier for the material. The backing readily separates from the adhesive prior to the application of the material to a surface.

Backslant Where a typeface can be made to slant backwards, in the opposite way to italic.

Back-up To print the reverse side of a sheet.

Bank A fine writing or typewriting paper, white or tinted, made in a range of weights from 45g/m^2 to under 63g/m^2. Heavier weights of otherwise similar material are termed 'bonds'.

BASIC **B**eginners **A**ll-purpose **S**ymbolic **I**nstruction **C**ode. The earliest, most popular language used on microcomputers.

Bed The base or table of a letterpress printing machine, in which the forme is locked in preparation for printing, or for cutting and creasing.

BPB **B**est **P**ractice **B**enchmarking. The process of comparing or 'benchmarking' an organisation against the counterpart in the best competing organisations. The object is to improve in all areas to become the best in all measured activities.

Bit **B**inary **i**nformation **t**ransfer' or 'binary digit'. The basic unit of information in computing and computer imagesetting; it represents a pulse, electrical charge or its absence. Each bit stands for one binary digit, 0 or 1. Bits are usually grouped together in blocks of eight to make *bytes*. Most computer operations work on byte-sized pieces of information.

Bitmap An image arranged according to bit location in columns. Resolution of a PostScript file processed through a RIP will have a bitmap image with the characteristics and resolution of the particular output device (for example, laser printer at 300 up to 1200dpi, imagesetters at 1270dpi up to 5080dpi).

Black box Colloquial term for a device or program which converts information from one form to another.

Blanket cylinder The cylinder on an offset lithographic printing machine on which the blanket (fabric coated with a rubber or synthetic compound) is carried and by means of which the printing image is taken from the plate and transferred to the substrate.

Bleed Printed matter which runs off the edge of the substrate; also used by bookbinders to describe over-cut margins and mutilated print.

Blind Term applied to a litho plate which has lost its image; also to book covers which are blocked or stamped without the use of ink or metallic effect.

Blind-blocking Blank impression made on book covers by binders' brass, without gold leaf, foil or ink.

Blister packaging Method of packaging in which an object is placed in a pre-formed, clear plastic tray, and backed by a printed card.

Block In binding, to impress or stamp a design upon the cover. The design can be blocked in coloured inks, gold leaf or metal foil (see *blind*). In printing, a letterpress block is the etched copper or zinc plate, mounted on wood or metal from which an illustration is printed.

Boards, bristol A fine quality of cardboard which may be made solid by pasting two or more sheets together.

Boards, chip Inexpensive board made from mechanical wood and waste materials: used unlined for binding cases, rigid boxes, show cards, and white lined for cartons.

Boards, mill A high-grade board, brown in colour, made from rope and other materials, which is very hard, tough and with a good finish. It is used for covers of better quality account and other books.

Boards, paste Board which contains two or more laminations of paper having a middle or lower quality.

Boards, pulp Manufactured from pulp as a homogeneous sheet on a cylinder machine, in a similar manner to paper.

Boards, straw A board made from straw, and used principally for making the covers of case books and cheap account books.

Body paper Paper forming the base of coated paper.

Bold A typeface that is heavier than normal weight, available in most type families.

Bolt Any folded edge of a section other than the binding fold.

Bond Similar to bank paper but heavier, usually supplied in $63g/m^2$ and over.

Bound book A book in which the boards of the cover have first been attached to it, the covering of leather, cloth, or other material being then affixed to the boards. Bound books are more expensive to produce and much stronger than cased books.

Boxhead ruling The space at the top of a ruled column for the insertion of printed or written headings for each column.

Broadsheet Any sheet in its basic size (not folded or cut).

Bromide A photographic paper used in graphic reproduction, phototypesetting and imagesetting on which a photographic image is created (see also *PMT*).

Buckram A binder's heavy cloth made from coarse textile thread and stiffened with size or glue. Very strong, wears well; used for account books when leather is too expensive.

Bulk Relative thickness of a sheet or sheets, for example, a bulky paper and a thin paper both of the same weight display different 'bulk'.

Bull's eye Printing defect caused by a dust particle holding the paper or board away from the printing surface.

Burst binding A type of adhesive binding in which the back of the book block is not sawn off, but is slit in places to allow glue to penetrate.

C sizes The C series within the ISO International paper sizes range which is mainly used for envelopes or folders suitable for enclosing stationery in the A sizes.

Calendar A machine used in the finishing operation of paper and board manufacture. It is composed of rollers between which the paper or board passes under pressure to give it a smooth finish.

Calibration bars These take the form of a strip of tones used to check printing quality throughout the process as a negative, positive, proof or printed copy. See *colour control bar*.

Caliper The thickness of a material.

Camera-ready artwork Finished artwork that is ready, without further preparation, to be photographed.

Carbonless paper Paper stock coated on the back and/or front with chemicals which react to form an image when written or typed on.

Carton A container generally made from paper/board, but sometimes partially or totally from plastic; delivered by the carton manufacturer to a user, either in flat or collapsed form, for assembly at the packaging point.

Cartridge A tough, opaque paper with a rough surface. Principally used for guard books, large envelopes, drawing and offset printing.

Case This represents the cover of a book prepared beforehand for affixing to the book.

Case binding The binding of printed books, which include leather, cloth and other forms of covering.

Character generation The projection of graphic images on the face of a cathode ray tube or similar display unit.

Chase A rectangular iron frame well below type height in which letterpress type and blocks, or cutting and creasing formes are locked in preparation for printing on the machine and certain other operations.

Chemical wood pulp Pulp that is prepared from chipped wood by treating with chemicals to remove the non-cellulose material. It is used in the better grade of wood pulp papers and boards, and improves the qualities of mechanical pulp when the two are mixed: also often referred to as *woodfree*.

Cheque paper Paper chemically treated in order to betray any tampering with the writing on the cheques.

China clay A fine white clay, used in papermaking for loading and coating.

Chip The basic building block for computers, made from silicon.

Choke A specific adjustment or distortion whereby the perimeter, in total or in part, of an element is slightly pulled in (*choked*) towards the centre of the element. Choking of an element is normally used in conjunction with the spreading (see *spread*) of a neighbouring element to ensure that colour registration standards are achieved. Choking of an element may be performed in a number of ways, manually, photomechanically or digitally, with various computer programs.

CMYK Initial letters indicating the printing 'subtractive' primary colours - **c**yan, **m**agenta, **y**ellow and blac**k**.

Coarse screen A halftone screen up to 35 lines per cm used in preparing illustrations for newsprint and similar surfaced papers.

Coated paper Paper which has received a coating on one or both sides. *Art papers* are coated papers, there are also *cast-coated*, which are high-gloss papers on which the coating has been allowed to harden in contact with a highly finished casting surface. In addition there are *brush-coated* papers; *chromo* papers which are clay-coated in a separate operation from papermaking; *roller-coated* papers; or *machine-coated* papers in which the paper is coated during the papermaking process.

Cold type Methods of typesetting by typewriter or early types of photosetting systems to produce copy suitable for reproduction or setting direct onto paper plates for offset litho printing.

Collate To check through the signatures or pagination of the sections of a book to ensure that they are complete and in correct sequence for binding. See *gathering*.

Collating marks Black step marks (usually 6-pt rule) printed on the back folds or sections and in progressively different positions so that any displacement of sections may be checked after gathering.

Colour control bar A coloured strip on the margin of the sheet which enables the platemaker and printer to check by eye or instrument the printing characteristics of each ink layer. See *calibration bar*.

Colour separation In photomechanical reproduction, the process of separating the various colours of a picture usually by colour filters or electronic scanning so that separate printing plates can be produced.

Colour work Printing more than one colour on a sheet, usually with some reference to register: also printing two or more partially-overlapping colours to obtain decorative or pictorial effect.

Contact screen Used to produce a halftone from continuous tone film or artwork using cameras or scanners.

Continuous tone Term often shortened to *contone*, it describes images which contain an apparently infinite range of shades and colours smoothly blended to provide a faithful reproduction of natural images.

Convertible press Type of sheet-fed press able to print either on one side of a sheet, or on both sides.

Covering The process by which a cover is affixed fully to the spine and both sides of a book.

CP/M Control Program for Microcomputers. A popular operating system for micros, word processors and front-end systems.

CPS Characters per second. This refers usually to the output speed of photo/ imagesetting.

CPU Central Processing Unit. In large computers, this may consist of a circuit that contains a number of chips, but for microcomputers the CPU is almost invariably a single chip, the microprocessor.

Crease To mechanically press a rule into heavy paper or board to enable folding without cracking. See *score*.

CRT Cathode Ray Tube; an electronic vacuum tube with a screen on which information (text or pictures) may be stored or displayed. They are used as displays in video display terminals, and to expose letter images onto film or paper in phototypesetters/imagesetters.

Cursor The moveable light spot on a VDU screen which allows the operator to identify a position on the display.

Cut-in index Style of index in which the divisions are cut into the edge of the book in steps: step index.

Daisy wheel A printing head used on typewriters and computer printers where individual characters are on the ends of 'petals'.

Dandy (Laid, spiral, wove), a cylinder of wire gauze on the papermaking machine which comes into contact with the paper while it is in a wet and elementary stage: the dandy roll impresses the watermark.

Database A collection of data items which are used frequently by programs. A database of any size would be kept on a disk, if not on several disks.

Deckle The width of web (machine width) which a papermaking machine is capable of making paper and board. This is limited by the deckle straps which were originally the movable wooden frame on the hand-mould used for papermaking.

Deckle edge The feathery edge occurring round the borders of a sheet of hand-made or mould-made paper, due to the deckle or frame of the mould: double deckle edged means two sides of a machine-made sheet are rough edged.

Densitometer A device for measuring the closeness of substance at a specific location on film or printed product, either by reflected or transmitted light. Densitometers vary in their sophistication and the number of features provided, such as colour, black-and-white, read-out memory, computer printout etc.

Descender Part of the lower-case letter below the x-height of the character as in g, q and p.

Diestamping An intaglio process of printing in which the resultant impression stands out in relief above the surface of the stamped material, either coloured (using inks) or blind (that is, without colour): relief stamping.

Diffusion transfer See *photomechanical transfer*.

Digital fount An electronically-stored type fount with characters stored as a series of digital signals.

Digital page composition DPC, also known as EPCS (*electronic page composition system*) or CEPS (*colour electronic* page system). A system designed to take a range of page elements (text, linework and images) and integrate them into a user-specified format. Image and text input to the system arrive on magnetic tape, by direct system interconnection or directly from an input scanning system.

Direct (work) Work which can be made from a (screen) negative obtained by photographing the original direct.

Direct screening A method of reproduction usually working via an enlarger in which colour copy is reproduced in the form of directly screened separations, by-passing the need for a separate screening operation.

Discretionary hyphen Keyboard hyphens which override the hyphenation programme on the computer.

Display matter Type displayed such as title pages, headings, jobbing work, as distinct from solid composition or body matter.

Display sizes of type Sizes of type from 14- to 72-pt, mainly used for display matter.

Distributing rollers The rollers on a printing machine which distribute the ink from the ink duct to the plate or forme inking rollers. They smooth out the ink film and should be arranged to prevent repeats or ghosting, which are printing defects where the image repeats or a 'ghost' or outline appears around the image areas.

Dot etching The process by which tone correction is applied to halftone negatives or positives.

Dot gain Enlargement of the halftone dot between film and print which should be assessed and allowed for in reproduction.

Dot matrix Imaging method in typewriters and computers. Each letter is made up of dots using a matrix of 5 x 7 or greater.

Drawn-on cover A paper book cover which is attached to the sewn book by gluing the spine.

Dry litho An offset litho process using a standard press, but with a plate which does not need damping to restrict the ink to the image area.

Dry offset See *letterset*.

Desktop publishing (DTP) A generic title given to the introduction of personal computers (PCs) to typesetting, page composition and image handling. The combination of all these gives electronic control within a single system of what was traditionally a specialist and segmented operation.

Dummy A sample of a proposed job made up with the actual materials and cut to the correct size to show bulk, style of binding, etc. Also a complete layout of a job showing position of type matter and illustrations, margins etc.

Duotone A two-colour halftone produced from two halftone images of the same original. Different visual effects can be obtained by using different screen angles, contrast ranges, special screens etc.

Duplex halftones Two-colour halftone plates made from a monochrome original, the second plate being used as a tint.

Duplex paper Paper of two qualities or colours which have been brought together and combined while in the wet state on the papermaking machine.

EBCDIC Extended Binary Coded Decimal Interchange Code.The IBM eight-bit standard character code.

Edges, sprinkled The spattering of book edges from a brush charged with liquid ink; used for decoration. May also be done with an air brush.

Edition A number of copies printed at any one time when some change has been made in type or format.

Electron beam See *radiation drying.*

Electrostatic printing A term used to describe where the printing plate, drum or belt is charged overall with electricity and light is reflected from the non-image areas of the original being copied, destroying the charges in these areas. Toner powder is then applied, which adheres only in the still-charged image area, fusing itself to the paper by heat. See also *laser printing.*

Embossing The process of raising, by an uninked block, letters or designs on card or strong paper.

Encoder A mechanism for converting data in one format to data in another, for example - RGB to CMYK.

End of line decisions Refers normally to hyphenation and justification, and decisions in photo-setting/imagesetting made by a keyboard operator or a computer.

Endpapers Lining sheets used at each end of a book, and used to attach the end sections to the cover.

EPS A file format, *Encapsulated PostScript,* used to transfer PostScript image information from one program to another.

Ethernet A networking system enabling the high-speed transfer of data between computer systems and peripherals over a co-axial link.

Exception dictionary Part of a computer's memory where words which do not hyphenate in line with standard rules may be stored.

Facsimile (fax) The transmission of copy, artwork or separations electronically from one location to another, to produce a duplicate (facsimile) of the original data.

Feint ruling Horizontal lines usually of pale blue ruled across account and manuscript books.

Filmsetting See *phototypesetting.*

Finishing This covers all operations after printing; also the hand operations of lettering and ornamenting the covers of a book.

First-and-third Description applied to a printed sheet where the printed matter appears on pages one and three when folded.

First-generation film An original plate-ready film, which has been exposed either through a camera or, more typically, through a laser imagesetter, and photographically processed.

First-generation photosetters Adaptions of hot-metal machines, for example, the early Monophoto, which set the letters by exposing through a photographic negative matrix based on the original Monotype grid.

Fit Proportion of space between two or more letters which can be modified (for example, tight fit) by adjusting the set-width.

Fixed space Space between words or characters not variable for justification purposes.

Flat back Bound sections having a square back, that is, not 'rounded and backed'.

Flat wire stitching See *stabbing*.

Flexography A relief process in which printing is done from rubber or plastic on a web-fed press using liquid inks.

Floppy disk A removable magnetic storage medium. Floppy disks come in a range of sizes, eg 8", 5.25" and 3.5". The most common used type is the 3.5", having considerably more storage capacity than the older 8" variety.

Flush A style of binding in which the covers and leaves are trimmed simultaneously as a final operation.

Flush left and right Type lining up vertically, either to left or right.

Folioing Numbering progressively by book openings instead of pages, the left and right pages have the same number.

Font American for fount. Both are pronounced 'font'. See *fount*.

Foredge The edge of a book opposite the binding edge, spine or back.

Formats Repetition of typographical or other commands, also text material which are determined by design and fed into the computer memory, allowing the keyboard operator to change typeface, measure, etc; also allowing repeated instructions to be done automatically. Formatting is used to program a computer by a single command to simplify repeated changes to text matter after it has been set and printed out - for example, the updating of a price list.

Forwarding In case binding, the processes involved in the making of a book after sewing and up to finishing.

Fount A set of type characters of the same design (and with hot metal, also the same size) for example, upper and lower case, numerals, punctuation marks, accents and ligatures.

Four-colour machine A printing machine which prints one side of the substrate in four colours as it passes through the machine.

Four-colour process Colour printing by means of the three subtractive primary colours (yellow, magenta, cyan) and black superimposed; the colours of the original having been separated by a photographic or electronic process.

Four-colour process inks Inks used for four-colour process printing (yellow, magenta, cyan and black).

French fold A sheet of paper with four pages printed on one side, and folded into four leaves without cutting the head. The inside four pages are then blank and printing appears on pages one, four, five and eight.

Front-end system The part of a computer or typesetting system responsible for the control of input, correction, manipulation and storage; as opposed to the 'back-end' which outputs the type/images.

Front lay See *lay*.

Fugitive colour Ink which is not stable when exposed to certain conditions of light, moisture, or atmosphere.

Full bound Style of binding in which the covering material is one piece of the same material - ie whole bound.

Function codes Codes which control setting system's functions, as opposed to input codes which produce characters.

Furnish The term used by the papermaker to describe the class and proportion of materials used in the making of paper.

g/m^2 Abbreviation of grams per square metre. A method of indicating the substance of paper or board (whatever the size of the paper/board or number of sheets in the package) on the basis of weight in grams per square metre.

Gathering To place in their correct order the sections or sheets to make up a book.

Generic mark-up A method of mark-up which describes the structure and other attributes of a document or print job in a rigorous and system-independent manner so that it can be processed for a number of different applications (see *SGML*).

Golf ball typewriter The type head used in some typewriters and strike-on typesetting equipment.

Grain (in paper) See *machine direction*.

Graining (in lithography) Roughening the surface of a metal printing plate by means of brushes, chemical reaction and abrasives so as to obtain a surface which will retain moisture and ink.

Grammage The weight of a material such as paper defined in g/m^2 (grams per square metre).

Graphics tablet Used in computing, composition and page make-up for layout or system control. A 'mouse', pen or puck on a drawing board controls movement on the video screen, traces in outlines, or permits selection of commands from a 'menu'.

Gravure printing A process in which the printing areas are below the non-printing surface. The recesses are filled with ink and the surplus is cleaned off the non-printing area with a blade before the paper contacts the whole surface and lifts the ink from the recesses.

Greaseproof A wood pulp paper which is made translucent by prolonged beating of the pulp.

Greyboards Case boards of a higher quality than chip boards; produced mainly in Holland.

Greyscale The depiction of grey tones between black and white. A greyscale monitor is able to display grey pixels as well as black and white, but not colour pixels.

Grid A regularly spaced set of lines in two dimensions to form a series of positional references. In electronic systems they may be used to position accurately text and image information for on-screen page layout.

Guard book A book with guards in the binding edge to prevent breaking of the back when filled with cuttings, samples, patterns etc.

Guards Strips of paper sewn between the leaves of a book, on which maps etc can be pasted.

Gutter The binding margin of a book.

H & J Hyphenation and justification.

Half-bound Style of binding in which the back and corner covering are of one material and the remainder of another.

Half-sheet work See *work-and-turn*.

Halftone screen Glass plate or film, cross-ruled with opaque lines and having transparent squares; used to split up the image into halftone dots (see *contact screen*).

Hard copy Typed or printed copy produced simultaneously with a tape or recording which allows operators to read and correct before photosetting or imagesetting.

Hard disk A fixed magnetic storage medium in which the data holding element cannot be removed. Hard disks have very large storage capacity, up to 1000Mb and above, enabling data to be rapidly accessed and manipulated.

Hard-sized (paper) A relative term applied to paper indicating a maximum of sizing. Lesser degrees are indicated by 'half-sized' and 'quarter-sized'.

Hardware The electronic components of a computer, as opposed to the programming *(software)*.

Headband Originally a narrow band of sewing round a strip of cane at the top and bottom of the spine of a hand-sewn book which adds strength to the binding. Imitation headbands made in long strips are sometimes glued to the head of a machine-sewn book in order to give it a better appearance.

Headline In composition, it takes the form of a line of type at the top of a page separated from the text by white space. A running headline is the title (or abbreviated title) of a book repeated at the top of every page of text, or at the top of left-hand *(verso)* pages, with the chapter headings of the contents of the two pages on the right-hand (recto) pages. In ruling, it takes the form of a single or double horizontal lines ruled across account books at the head and from which down lines drop.

Headliner machine See *photolettering*.

Heat-set drying Drying a web or sheet of paper or board by passing it through a drying unit which forms part of the machine. Special heat-setting inks have to be used.

Hicky See *bull's eye*.

Highlight The whitest part of a halftone when printed.

Hollow The space in the back of the book between the two boards of the cover.

Hot-foil A printing technique using very thin aluminium foil in a variety of metallic colours, such as gold, silver, red and blue. The metallic foil is released from the carrier base onto a substrate by the application of heat and pressure from a metal printing plate which bears the image to be hot-foiled.

House corrections Corrections in galley or page proofs, other than those made by the author.

Hue The colour defining component of a point in an image. *Hue* combined with *saturation* fully defines a colour.

Icons Symbols used to make explanations shorter, relating to the manipulation of specific menus on screen: the Apple Mac system is a good example of the use of icons.

Image The ink-carrying areas of a lithographic printing surface plate.

Image master Photographic original for founts of typefaces used in photosetting.

Imagesetter An output typesetting system which has the ability to combine all the elements of a page - text, tints, graphics, etc - directly onto paper/bromide, film or polyester plate material.

Impose To plan film or pages prior to platemaking.

Imposition schemes Plans for the arrangement of the pages of a book so that they will follow in the correct sequence when folded.

Indented A shortened line of type set over to the right of the normal margin.

India paper A very thin opaque rag paper used for books when extreme lightness or thinness is desired: originally imported from China.

Indirect (work) Work which cannot be made from a screen negative direct, but for which a continuous tone negative and positive must be made before the screen negative can be produced.

Inferior characters and figures Letters or numbers which are smaller than the text size and are positioned on or below the baseline, also known as *subscript*.

Infra-red See *radiation drying*.

Ink jet A non-impact printing process in which droplets of ink are projected onto paper or other material, in a computer-determined pattern.

Inner An imposition containing the pages which fall on the inside of a printed sheet in sheetwork - the reverse of the *outer forme*.

Input Data for processing by a computer prior to outputting.

Input device Any device that can apply an input to the computer. This includes the keyboard, disk drive, tape unit, voice recognition and any other peripherals that supply input signals.

Insert A piece of paper or card laid between the leaves of a book and not secured in any way.

Insetting Placing one section inside another resulting in insetted work.

Interactive Term which relates to the situation when the operator working on a VDU or system can see what is happening and acts immediately to alter anything as required.

Interface A general term used to describe the method by which two independent systems may communicate with each other. The term interface is usually used in reference to electronic system interconnection, but may also refer to the way in which users relate to the equipment they operate.

Interleaving In printing, the placing of sheets of paper between printed sheets as they come from the machine to prevent set-off; also known as *slip sheeting*. In bookbinding it covers insetting into and folding around the sections of a book paper different from that used in the general body of the book, such as writing paper and blotting paper: also the alternating of processed and plain sheets, for example, in a duplicate book.

International paper sizes The standard range of metric paper sizes as per definition of the International Standards Organisation (ISO) and British Standards Institution.

Ion deposition Non-impact printing process in which ions are projected from a replaceable print cartridge onto a rotating drum to form a latent dot matrix image.

Justification The even and equal spacing of words and blocks to a predetermined measure: 'to justify a line' is to space out a line of type to the required measure.

Kerning The process of altering the space in between type characters to achieve a more aesthetically pleasing arrangement of the letters of a word.

Kettle-stitch In binding, the stitch at the top and bottom of the spine which connects each signature to the following one.

Key The outline of a drawing which is transferred or used as a guide in the production of printing plates so that the various colours will register with each other; also relates to the design which acts as the guide for position and registration of the other colours. Further examples of the term include the character key on a typesetting keyboard and to set via a keyboard.

Knocking-up To make the edges of a pile of paper or board straight, regular or flush.

Laminating The application of transparent plastic film, usually with a high-gloss finish, to the surface of printed matter to enhance its appearance and to increase its durability.

Landscape Oblong loose or folded printed sheet, or book, having its long sides at head and foot.

Laser **L**ight **A**mplification by **S**timulated **E**mission of **R**adiation - a fine beam of light, sometimes with considerable energy, used in imagesetting, colour scanning, copy scanning, platemaking, engraving, and cutting and creasing forme-making.

Laser engraving The process of engraving an image onto a printing plate or, more typically, a printing cylinder coated with rubber, using an intense laser beam. It is used for continuous patterns where a conventional printing plate join would be revealed in the printed image.

Laser imagesetter/ recorder/plotter Descriptions identifying the film/plate material output devices of various manufacturers. Film exposed at high resolution using laser light source direct from digital information stored on hard disk. The units have job-index and job-queuing programs as well as manual or automatic film-loading/unloading facilities.

Laser printing A form of electrostatic printing in which the image is not created by reflection from an original (electrostatic copying) but by switching a laser on and off according to digital information from a computer.

Lay The position of the print on a sheet of paper or board. Lays (front and side) - the guides or gauges (at front and side) to which paper or board is fed before being printed or otherwise processed on a machine eg folding. *Lay edges* - the edges of a sheet which are laid against the front and side lays.

Lead Term covering strips of metal less than type-high, used as general spacing material (thickness: 1-pt, 1.5-pt [thin], 2-pt [middle], 3-pt [thick]; also to lead, in all composition, to add space between lines of type. The term comes from hot-metal letterpress printing, in modern setting systems it is more often referred to as *interlinear spacing*.

Leader A type character having two, three or four dots in line, used to guide the eye across a space or other relevant matter, as in tables.

Leaf A sheet of a book, containing two pages one on each side. Thus a section of a book containing 32 leaves has 64 pages.

Letterpress printing A process in which the printing surface of metal, plastic, photopolymer or rubber is raised above the non-printing surface. The ink rollers and the substrate touch only the relief printing surface.

Letterset Offset letterpress printing, using a wrap-round relief plate on a litho press; also called *dry offset*.

Letterspacing To increase the standard space between characters to fill a line or enhance the visual look of the words. See *kerning*.

Ligature Two or more letters joined together, and forming one type character as fi, fl, ff, ffi, ffl.

Light emitting diode (LED) A semi-conductor that produces a light when a voltage is applied. In small sizes used in photocomposition/imagesetting and colour scanning.

Light pen Light-sensitive stylus used with certain VDUs for design or editing.

Limp cover A flexible book cover, as distinct from a stiff board cover.

Line block A relief block produced from a line drawing and without the use of a halftone screen.

Line caster Hot-metal typesetting machine that produces lines or slugs of type.

Line drawing A typical drawing which would be produced with pen and brush when using a full charge of ink, thus making lines of comparable photographic value.

Line feed The distance, usually in points, between base lines or successive text lines in phototypesetting/imagesetting.

Line printer High-speed, tape or computer-activated machine producing typewriter-like printouts.

Linen finish A surface impressed on paper or board to make it resemble linen, usually produced by passing the web between engraved cylinders. Similar and various other patterns can be given to paper or board after printing.

Lining (second) After the first lining of mull has been placed on the back of the book, a sheet lining, often paper, is glued on to strengthen the book.

Lithographic printing A process in which the printing and non-printing surfaces are on the same plane and the substrate makes contact with the whole surface. The printing part of the surface is treated to receive and transmit ink to the paper, usually via a blanket (see *offset printing*), the non-printing surface is treated to attract water and thus rejects ink from the ink roller, which touches the whole surface.

Loading Clay or other mineral included in the furnish of a paper or board to produce a more solid (opaque) and smoother sheet, also known as *fillers*.

Logo An image or symbol constructed from shapes, designs and letters, designed to represent an organisation, trademark, etc.

Look-through The appearance of paper or board when held up against a strong light.

Lower case Term covering small letters of the alphabet as distinct from capitals; also the divided wooden tray which holds the lower-case sorts, as used for hot-metal founders typesetting.

Machine direction The long way of the paper web (or board) and the direction in which the cellulose fibres tend to lie due to the motion of the papermaking machine. The sheet has stronger physical properties in the machine direction and shows less dimensional variation when subjected to changes in humidity. Also, the direction in which a product is printed in a reel-fed printing machine - for example, the head of the label first along the web, the foot of the label first along the web, wide edge of the label leading along the web, narrow edge of the label leading along the web.

Manifold A thin, strong, smooth surface paper used for duplicating or copying, of substance under 45g/m^2.

Matrix, also matrice In typefounding, a copper mould from which a typeface is cast. In typesetting machines a recessed image in a brass plate which is mechanically set and justified and from which a slug is made.

MRP **M**aterial **R**equirement **P**lanning. A system used, normally in conjunction with a computer, in calculating the material required over a period, based on the stock-in-hand and the existing/projected orders for that period.

MRP II **M**anufacturing **R**esource **P**lanning **II**. This takes the form of capacity planning. It is based on the MRP database plus the sequence of all operations in the organisation and the corresponding times for each activity, alongside the capacity of every department.

Matt art An art paper or board with a dull eggshell finish.

Mechanical printing Any paper containing a proportion of mechanical wood pulp.

Mechanical wood pulp Produced by grinding wood mechanically; used in cheap papers, such as newsprint, and combined with larger proportions of chemical wood pulp for better qualities.

Menu The choice of operations displayed on a VDU.

Merge A method of combining matter on two or more tapes or disks into one, using a computer to incorporate amendments or new copy into existing copy and to produce a clean tape or disk for typesetting.

Mesh (screen printing) The weave dimension and angle of the fabric of material used for preparing silk screen stencils.

MF (paper) Abbreviation of *machine-finished* or *mill-finished*; paper finished on the papermaking machine but not super-calendered.

MG (paper) Abbreviation of *mill-glazed* or *machine-glazed*; applied to a large range of papers which are characteristically rough on one side and highly glazed on the other.

MICR **M**agnetic **I**nk **C**haracter **R**ecognition. Automatic sorting method used, for example, on cheques, based on the printing of numbers in magnetic ink.

Microcomputer A small computer, usually made to sit on a desk top.

Microfiche A sheet of film, typically 150 x 150mm, holding in reduced size, many pages of larger documents.

Microprocessor The 'chip' that forms the central processing unit of a computer.

Millboard A high grade board, brown in colour, made from rope and other materials; very hard and tough with a good finish which is used for covers of better quality case-bound books.

Mill-finished (paper) See *MF*.

Mill-glazed (paper) See *MG*.

Mixing Having more than one typeface, style or size in a line of text.

Modem **Mo**dulator/**dem**odulator. A device wired to a computer which translates signals to enable them to be transmitted on telephone lines.

Modification With digital imagesetters/photosetters and display setters, characters can be condensed, expanded or italicised by digital or optical means.

Moiré pattern In colour printing the term describes an irregular and unwanted conglomeration of screen dots of the different printing inks, which cause disturbing patterns or patches, either over the whole image in certain combinations. Moiré is mainly caused by incorrect screen-angling.

Montage Term used in the graphic industry for a number of operations: *photomontage* - combination and often blending of images; *montage of pages* - page make-up; *montage of film* - mounting several colour separation films of one printing colour in register for subsequent transfer to the printing plate.

Mouse Electronic device, used on a graphics tablet, for drawing or 'pointing' to certain areas of the screen on a computer.

Mull An open net fabric which is fixed to the backs of case-bound books, slightly overlapping front and back cover boards, to give strength.

No-flash In phototypesetting, a command to omit exposure (and the letters/words involved) from output.

Non-counting keyboard A keyboard used to type copy for setting, with the operator producing a continuous stream of characters, which is fed to the computer to determine justification and hyphenation.

Numbering-at-press To number a job on the printing machine by means of numbering boxes.

Oblong See *landscape*.

OCR **O**ptical **C**haracter **R**ecognition - typewritten or printed matter capable of being read opto-electronically using a scanner, for subsequent imagesetting/phototypesetting.

Off-line Not connected - eg a computer printer may be disabled by switching it off-line. Another use is where information entered into a computer is processed at a later time, without the operator being present. Many EPC and graphic systems manipulate the actual data after the job to be done has been set up by an operator. The off-line component of the job often takes much longer than the initial set-up. This may be referred to as post processing and forms a major limiting factor to the efficiency of such systems.

Offset printing A lithographic method of printing in which the ink is first transferred from the image to an offset blanket and then to the stock which may be paper, card, metal or other material.

One-letter index An index consisting of 24 letters or divisions, omitting x and z from the alphabet.

On-line The opposite of off-line - that is, the transfer of data from one device to another is done via a direct link. On-line connections require that the linked devices understand their data formats and structures and are able to keep track with the speed of data transfer.

Original The term applied to copy which is to be reproduced.

Original plate A letterpress block or relief printing plate produced by a block maker, or by a photomechanical etching process, as distinct from electrotyping or stereotyping.

Orthochromatic A photographic film insensitive to red light; used for monochrome reproduction or on scanners using non-red lasers.

Outer An imposition containing the first and last pages of a printed sheet in sheetwork; as distinct from *inner forme*.

Overlap (cover) A cover of a paper-bound book which extends beyond the edges of the pages of the book.

Over-running To turn over words from one line to the next for several successive lines after a deletion or insertion.

Overs The quantity of unit production, for example, books and sheets, delivered to the customer above the net amount ordered, usually charged at a run-on rate.

Paged A book is said to be paged when the pages are numbered consecutively; as distinct from folioed.

Palette A range of colours, accessed on electronic systems from a colour data base and displayed on the screen, used for a specific job. The palette which may consist of colours classified according to the *Pantone Matching System* or other colour systems, may be updated and changed in seconds.

Panchromatic A photographic film or plate sensitive to all visible colours of the spectrum.

Pantone *Pantone, Pantone Matching System* and *PMS* + are Pantone Inc's check-standard trademarks for colour standards, colour data, colour reproduction and colour reproduction materials, and other colour-related products and services, meeting its specifications, control and quality requirements.

Paste board Board made of two or more laminations of paper or board.

Paste-up Any matter pasted up as copy for photographic reproduction.

PDL Page Description Language, such as *PostScript*, identifies the parameters of a page as a set of co-ordinates and with the use of a compatible raster image processor translates the data into a suitable form for outputting. See *Postscript*.

Perfect binding See *adhesive binding*.

Perfected sheet A sheet printed on both sides.

Perfecting Printing the second side of a sheet; backing-up.

Perfector (machine) A printing machine which prints both sides of the sheet as it passes through the machine.

Perforating-at-press To perforate a job on the printing machine by means of a perforating rule.

Photolettering Generally a method of photographically producing display setting of typefaces from film founts in machines of varying complexity, but not so automated as text phototypesetters.

Photolithography The process of reproducing an image on metal by photography for lithographic printing.

Photomechanical transfer (PMT) A method by which an image is photographed and screened on to a paper negative which by chemical transfer produces a bromide print. This may be reproduced dot-for-dot in platemaking, or pasted-up with unscreened text for further reproduction.

Phototypesetting The setting of type matter on film or photographic paper, also known as photosetting.

Pi characters Characters omitted from a normal type fount or master, for example, accents, mathematical signs, but included on a separate fount.

Pigment Particles that absorb and reflect light and appear coloured to the eye; also refers to the substance that give ink its colour.

Pixel From **Pic(x)**ture **el**ement, the smallest part of a picture on a computer screen.

Plate Any relief, planographic, or intaglio surface; also an illustration of a book printed separately from the text and usually on different paper.

Plate cylinder The cylindrical surface on a rotary printing press, which carries the printing surface.

Plate guarded and hooked A plate secured into a book by means of a narrow strip of paper or linen (guard) pasted on to its back edge, and the guard folded in or round a section and sewn with the section.

Plate hooked on own guard A plate secured into a book by folding the margin of the back edge in or round a section and sewing it with the section.

Platen (machine) A small direct impression letterpress printing machine, sometimes termed a *jobbing platen*.

Plates guarded and joined Two plates joined together by means of a strip of paper or linen, thus forming four pages which can be included in the sewing of the sections of the book.

Plate hooked on own guard Two plates joined together by means of a narrow strip of paste down the back edge of one, so that they can be folded to form a four-page section which can be included in the sewing of the sections of a book.

PMT See *photomechanical transfer*.

Point system The use of a typographic standard 12-pt pica or 4.23mm to which all other measurements are referred.

PostScript A Page Description Language (PDL) developed by Adobe, which describes the contents and layout of a page. PostScript also serves as a programing language whereby the PostScript code is executed by PostScript RIP in the output device in order to produce a printout or film containing the page.

Pre-sensitised plate A printing plate precoated for direct exposure, made in positive or negative form.

Printer The unit that prints out information from a computer which can take the form of a daisy wheel, dot matrix, laser, ink jet or thermographic device.

Printing cylinder See *plate cylinder*.

Process colours The printer's traditional subtractive primary colours: cyan, magenta, yellow and black.

Program The instructions *(software)* that enable a computer to carry out the tasks desired.

Progressive proofs A set of proofs showing each plate of a set printed in its appropriate colour and in registered combination to act as a guide for the printer.

Proof A version of a document or colour illustration produced specifically for the purpose of review prior to reproduction.

Quad left, right or centre Term used to describe commands to make lines flush left, right or centred.

Quality The totality of features and characteristics of a product or service that bear on its ability to satisfy a given need or requirement; also described as 'fitness for purpose' or value for money as perceived by the customer.

Quality assurance All activities and functions concerned with the attainment of quality.

Quality control The operational techniques and activities which sustain the product or service quality to specified requirements.

Quality system A written description of every activity within the organisation which can directly affect the quality of the output produced or processed.

Quarter bound Style of binding in which the back covering is of one material and the sides of another.

Qwerty Standard typewriter keyboard layout used in the printing industry; the term also applies to the arrangement of keys on the upper left-hand of the board.

Radiation drying Accelerated drying of specially formulated inks and varnishes by infra-red (IR), ultra-violet (UV) or electron beam radiation.

Ragged right Term used to describe command to use a fixed word space, not allowing type to line vertically on the right; also known as *unjustified*.

Raster The method used in most imagesetters and VDUs to 'draw' the image, each image being made of a series of parallel, or rastered, lines which are switched on and off as they cross the image area. The alternative method is to use *vectors*.

Register The printing of two or more plates in juxtaposition so that they complete a design if printed on the same side of the sheet or back up accurately if printed on opposite sides of the sheet.

Register marks Marks placed in the same relative position on sets of printing plates so that when the marks are superimposed in printing the work falls into correct position.

Retouching The treatment of a photographic negative or positive manipulation to modify tonal values or to compensate for imperfections.

Reverse leading A setting function allowing the film/paper to be moved in the opposite direction to normal, thus achieving typographical effects, such as in multicolumn work or maths setting.

Reversing Altering the original from left to right in the reproduction and vice versa.

RGB The abbreviation of red, green, blue (the additive primary colours) as opposed to Y, M, C and (K), (the subtractive primary colours).

Right reading Paper/film, positive/negative from a image/phototypesetter which can be read in the usual way, that is - left to right.

RIP Acronym for **R**aster **I**mage **P**rocessor. PC-based graphic workstations usually produce files in a very compact form based on vector definitions. However, these are not directly suitable for output as all plotters and scanning systems need raster data to operate. RIP technology provides the link between *vector* and *raster* systems. PostScript is an example of a vector data generator.

Rosette The pattern created when all four-colour halftone screens are placed at the traditional angles.

Rounding and backing The hand or machine operation of shaping a book after sewing so that the back is convex and the foredge concave, and the formation of a shoulder against which to fit the cover boards.

Run-through A term in ruling where the lines run from one edge of the paper to the opposing edge without a break.

Saddle-wire stitching To stitch with wire through the back of folded work.

SC (paper) Abbreviation of **s**uper-**c**alendered.

Scanner Electronic colour scanners produce, from colour transparencies or colour copy, colour corrected screened separations for the four printing ink colours.

Score To partially cut/crease with a rule into heavy paper or board to break the grain and so enable easier folding.

Screen printing Often called *silk screen printing* from the material formerly used for the screen. A stencil process with the printing and non-printing areas on one surface. The printing (image) area is open and produced by various forms of stencil. The substrate is placed under the screen and ink is passed across the top of the screen and forced through the open (printing) areas on to the substrate below.

Scrolling This is a technique used on a VDU to recall information from the display memory: as each line is recalled, all existing lines move on the screen up or down by one line to make room for the next line.

Section A folded sheet of paper forming part of a book; sections are sometimes made of insetted folded sheets of four, eight, 16 or more pages.

Separation Term used in the reprographic industry to describe the films which represent the yellow, magenta, cyan and black content of an image: by printing these four separations one on top of the other, most of the mixed colours of the image can be regenerated in the printing process.

Set The width of a type character. In the Monotype unit system, the width (set) of the widest character (em) of a fount is measured in points and sub-divided into units, which are one eighteenth of the set. All characters are multiples of units of their own set and their location in the matrix case is determined accordingly. In modern image/photosetting systems the units are finer eg Monotype units are one ninety-sixth of the set.

Set-off The marking of the underside of a printed sheet by the transfer of ink from the sheet on which it lays.

Sew To fasten the sections of a book together by passing thread through the centre fold of each section in such a way as to secure it to the slips: in distinction from *stitch*.

SGML **S**tandard **G**eneric (or Generalised) **M**ark-up **L**anguage. A versatile code used to mark-up and identify the various elements of a document for outputting in photosetting or other form. See *ASPIC*.

Sheet-fed rotary A printing machine on which the printing surface is fixed around a cylinder.

Sheet work A certain number of pages are imposed in two formes, one printed on one side and the other on the reverse side (backing up) - *inner* and *outer* - each backed-up sheet producing one perfect copy; also sometimes known as *work-and-back*.

Shrink wrapping Method of packing printed products by surrounding them with plastic, then shrinking by heat.

Side lay See *lay*.

Side notes Short lines of matter set in the margins.

Side stitching To stitch through the side from front to back at the binding edge with thread or wire. See *stabbing*.

Signature The consecutive number or letter which is printed at the foot of the first page of a section to enable a binder to check the position and completeness of the sections. Signatures are often indicated by printing a rule in the back of each section so that when the sections are folded and gathered the signatures appear 'stepped' on the back fold.

Size Resin or other sizing material included in the furnish of a paper to bind the fibres and loading together and to provide greater resistance to ink and greater strength in the sheet.

Skin packaging Method of packaging by which thin, clear plastic is shrunk onto an object backed by a printed card.

Skiver A cheap leather made of split skins; also the outer or grain side of such leather.

Slitting Term which covers cutting a sheet or web into two, or more parts, after it is printed and before it is delivered.

Slot Any pattern of hole, other than round, punched in paper or board.

Slug A complete line of type, as produced by a linecaster's machine.

Software The programs that enable the computer to perform its tasks.

Solid Type set without leads (as in hot metal setting) or additional feed between the lines - eg 10 on 10pt or 11 on 11pt.

Spaces Metal blanks less than type-height used for spaces between words or letters such as hair, thin, middle, thick, as used in hot metal setting; term also used for the same purpose in image/phototypesetting.

Spine See *back*.

Spoilage Term covering unprofitable materials and labour, the cost of which cannot be charged to a specific customer.

Spread The process generally carried out enlarging the width of line work. The inverse function, *choke*, is used to reduce the width of line work by using the same process but from the positive image. This is done to ensure that there is no gap between the linework and the surround area. The line work now larger, spreads over from its original area to give an overlay, simplifies the printing process by reducing the need for absolute accuracy of the press; also helping to compensate for shrinkage and stretch in the substrate to be printed. See *choke*.

Spring back Pieces of strawboard or millboard rolled to the shape of the back.

Square back See *flat back*.

Squares Protective projections of the cover of a book beyond the edges of the leaves.

Stabbing To stitch with wire through the side of gathered work at the binding edge.

Standing matter Composed matter for letterpress printing kept in chase or stores pending possible reprint.

Step index See *cut-in index*.

Stitch To sew, staple or otherwise fasten together by means of thread or wire the leaves or signatures of a book or pamphlet. The different styles of stitching are: *double stitch*, where two loops of a single thread are fastened in the centre of the fold; *machine stitch*, where a lock stitch is made; *saddle* or *saddle-back stitch*, where the centre of the fold is placed across the saddle in the machine and wire staples are driven through and clenched on the inside; *side stitch*, where the thread or wire is stitched through the side of the fold; *single stitch*, where a single loop is drawn through the centre and tied; *wire stitch*, in which staples are made, inserted and clenched by a machine from a continuous piece of wire, as in the saddleback stitch: as distinct from *sew*.

Strike-on See *cold type*.

Stringing To insert and tie string on hanging cards, catalogues, and other work either singly or in batches.

Strip gumming To apply, by hand or machine, water-soluble gum to paper in strips and then to dry.

Stripping Term used to glue a strip of cloth or paper to the back of a paperbound book or pad as a reinforcement; also to remove the waste material from between cartons and other shaped work.

Substance See *grammage*.

Sulphite Wood pulp prepared by the sulphite process. Sulphate wood pulp is prepared with sulphate of soda, caustic soda and sulphite of soda.

Super-calendered (SC) Paper which has been given a smooth glazed surface by passing between the calender rolls under heavy pressure.

Superior characters and figures Letters or numbers which are smaller than text size and are positioned above the top of the body.

Taping Pasted strips of linen, calico or other suitable material attached to inside or outside of sections to strengthen the paper, usually 10-12mm wide; also between sections to prevent breaking away.

Thermal printing Non-impact printing process in which heat is transferred from a digitally-controlled print head to a substrate causing a change in colour.

Thick space A type space having a width of one-third of its own body.

Thin space A type space having a width of one-fifth of its own body.

Thumb (index) Style of index where the divisions are cut into the edge of the book but not 'stepped'; as distinct from cut-in and tab index.

TIFF Tagged Image File Format, a file format for exchanging bitmapped images (usually scans) between applications.

Tint laying Term used to cover preparing the many patterns of mechanical shading.

Tints Mechanical shading in line areas, normally available in 5% steps from 5% to 95%.

Titling (founts) Any fount of capitals, generally full-faced and used for headlines or titles.

TQM Total Quality Management. A quality process which involves everyone in an organisation working together to the common goals of improving quality, customer satisfaction, eliminating errors and waste - the introduction of TQM often follows on after an organisation has introduced BS5750/ISO9000 or at the time of working towards certification.

Turned-in When the material used on the cover of a book is turned-in round the edges, so as not to leave the edges of the cover boards exposed, the cover is termed 'turned in'. See *flush*.

Twin-wire paper Even-sided paper produced from two webs joined together while still wet with their undersides at the centre.

Two-colour machine A printing machine which prints one side of the sheet in two colours as it passes through the machine.

Typeface classification British Standard 2961:1967 classifies typefaces by characteristics and style and not by origin; many other classifications exist.

Type-height A letterpress printing plate is said to be type-high when it is mounted to the correct height for machine printing.

Ultra-violet UV. See *radiation drying*.

Under-colour removal UCR. In the four-colour printing process, removal of part of the cyan, magenta and yellow, while adding extra black: its use leads to the overall reduction of the total quantity of ink used.

Underside (of a sheet) The surface of the web of paper which receives the impression of the machine wire on the papermaking machine.

Unit system Term based on the division of the em of the body size, each character of a fount having its own unit value. These systems, which vary according to manufacturer, are essential for justification and tabulation of lines. See *set*.

Unjustified Text setting where lines of type align vertically on one side, while being ragged on the other: wordspacing is kept to a constant value.

Unsewn binding See *adhesive binding*.

Upper case Term for capital letters (caps), also the type case which held the capital letters.

Variable space The space inserted between words to spread and justify the line to the required measure.

Varnishing To apply oil, synthetic, spirit, cellulose or water varnish to printed matter by hand or machine to enhance its appearance or to increase its durability.

VDU/VDT Visual display unit/visual display terminal. A display unit which consists of a cathode ray tube on which characters may be displayed, representing data read from the memory of a computer. The unit also incorporates a keyboard on-line to the computer to manipulate data within it.

Vegetable parchment A greaseproof paper, usually thicker and of better quality than paper termed 'greaseproof'.

Vehicle (of ink) Medium or varnish in which the pigment of a printing ink is carried or suspended.

Vellum finish A finish applied to paper and smoother than parchment.

Visual The design concept drawn, either manually or electronically, in colour to provide an impression of the final image.

Volume basis Term used mainly in book printing which denotes the thickness (bulk) of 100 sheets of a given paper in 100g/m^2.

Web Paper or board when made is wound on a roll or web. 'In the direction of the web' means in the direction of the run of the paper-making machine when the substrate is made. The direction of the web is important in work printed to register, as paper and board stretches more across the web than in the direction of the web.

Web-fed Presses which are fed by paper from a reel as distinct from separate sheets.

Web offset Reel-fed offset litho printing. Three main systems of presses exist *blanket-to-blanket* in which two plate and two blanket cylinders per unit print and perfect the web of paper or board; *three-cylinder system* in which plate, blanket and impression cylinders operate in the usual manner to print one side of the paper or board; and *satellite* or *planetary system* in which two, three or four plate and blanket cylinders are arranged around common impression cylinders to print one side of the web in several colours.

White-out Term covering to space out composed matter to fill the allotted space, or to improve the typographical effect; also to paint out matter not required for reproduction on artwork.

Whole or full-bound Style of binding in which the covering material is of one piece throughout.

Wire-mark The impression of the machine wire imparted to the underside of the web of paper on a papermaking machine.

Woodfree paper Any paper made from chemical wood pulp and containing no mechanical wood pulp - see *chemical wood pulp*.

Word processing The input, editing, organisation and storage of words/data using computer-based equipment.

Work-and-back See *sheetwork*.

Work-and-turn When matter is printed in its entirety on both sides of a sheet by using the same gripper edge.

WORM Write Once Read Many times. This describes a digital storage medium (usually optical) to which you may send information - eg an image. This image is then stored permanently on the disk which cannot be erased or altered, but can be read many times. It is a method of storage which is extremely useful for archive purposes.

Wove Paper which shows an even texture rather than a parallel line pattern.

Wrappering The process of attaching a paper or board cover by means of a strip of glue at the spine of gathered work (stabbed or sewn).

WYSIWYG What You See Is What You Get. An acronym used to describe a visual display showing a representative replica of the resultant output in paper or film form.

Xerography Proprietary name for a form of electrostatic printing.

x-height The height of lower case letters having neither ascenders nor descenders, as x, m and u.

Bibliography

The books and other literature listed here are useful reference material for enlarging on subjects covered within the text book.

Current information can also be found in the wide range of trade, management and business press publications.

Business, commercial and management

Buying a printer's Management Information System
(BPIF, 1993)

Customs of the trade for the manufacture of books
(BPIF, 1992)

Customs of the trade for the manufacture of cartons
(BPIF, 1990)

Customs of the trade for the production of periodicals
(BPIF, 1987)

Estimating for printers
(BPIF, 1989)

Getting the measure of your business
(BPIF, 1989)

Handbook of management
edited by D Lock (Gower Publishing Co Ltd, Aldershot - third edition, 1992)

ISO 9000
B Rothery (Gower Publishing Co Ltd, Aldershot 1992)

Law for printers
(BPIF, 1992)

Managing information
D A Wilson (Butterworth-Heinemann Ltd, Oxford, 1993)

Quality is free
P B Crosby (New American Library, New York, 1980)

Quality: sustaining customer service
L K Taylor (Century Business, London, 1993)

Right every time
F Price (Gower Publishing Co Ltd, Aldershot, 1990)

The management task
R Dixon (Butterworth - Heinemann Ltd, Oxford, 1993)

The power of added value
D Chapman and B Hill (Eaglehead Publishing Ltd, Cirencester, 1994)

Total quality management
K Holmes (PIRA, 1992)

Total quality management
J S Oakland (Butterworth Heinemann, Oxford, 1992)

VAT on printing
(BPIF, 1990)

Technical

Colour Concepts series
(Du Pont [UK] Ltd)

Introduction to printing technology
(BPIF, 1992)

Pocket pal - a graphic arts production handbook
(International Paper Co, Memphis, 1989)

Standard folding impositions
(BPIF, 1992)

The print and production manual
(Blueprint Publishing, London, 1989)

Troubleshooting for printers
(BPIF, 1993)

General reference - *British Standards Institution*

BS 1413	Page sizes for books
BS 4000	Sizes of papers and boards - (ISO 478, 479, 593)
BS 4778	Quality vocabulary
BS 5750	Quality systems: Part 0 - principal concepts and applications
BS 5750	Part 0: Section 0.1 (ISO 9000) - guide to the selection and use
BS 5750	Part 0: Section 0.2 (ISO 9004) - guide to quality management and quality systems elements
BS 5750	Quality systems: Part 2 (ISO 9002) - specification for production and installation
BS 5750	Part 4 - guide to the use of BS 5750
BS 5750	Part 8 - guide to quality management and quality systems elements for services
BS 7229	Guide to quality systems auditing

Index

D

E

F

G

I

J

L

M

N

O

P

U

V

W